CHEMICAL
ACHIEVERS

Left, top to bottom: Rachel Carson (Courtesy Beinecke Rare Book and Manuscript Library, Yale University Library); Robert H. Richards and Ellen Swallow Richards (Courtesy MIT Museum); Joseph Louis Gay-Lussac and Jean-Baptiste Biot in their balloon (From Louis Figuier, *Les Merveilles de la Science* [4 vols.; Paris, 1867–70], *2*, p. 537). **Right, top to bottom:** Herbert Blades, Paul Morgan, Stephanie Kwolek, John Griffing, and Eugene Magat (Courtesy Eugene Magat); Walter Lincoln Hawkins (Courtesy AT&T Archives); Wallace Hume Carothers (From the Carl Marvel papers).

CHEMICAL ACHIEVERS

THE HUMAN FACE OF THE CHEMICAL SCIENCES

MARY ELLEN BOWDEN

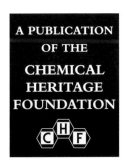

A PUBLICATION OF THE CHEMICAL HERITAGE FOUNDATION

9625903

DIRECTOR OF PUBLICATIONS: Frances Coulborn Kohler
PRODUCTION EDITOR: Patricia Wieland

Printed in the United States of America.

BOOK DESIGNED BY Sylvia Barkan
COMPOSITION BY Patricia Wieland
PRINTED BY Thomson-Shore, Inc.

For information about CHF publications write
Chemical Heritage Foundation
315 Chestnut Street
Philadelphia, PA 19106-2702, USA
Fax: (215) 925-1954
Website: http://www.chemheritage.org

Library of Congress Cataloging-in-Publication Data
Mary Ellen Bowden.
    Chemical achievers : the human face of chemistry / Mary Ellen Bowden.
        p.   cm.
    Includes bibliographical references and index.
    ISBN 0-941901-12-2 (paper : alk. paper)
    1. Chemists—Biography. 2. Chemistry—History. I. Title.
QD21.B67   1997
540'.92'2—dc21
[B]   97-5508
                                                          CIP

∞ The paper used in this publication meets the minimum requirements of
the American National Standard for Information Sciences—Permanence of
Paper for Printed Library Materials, ANSI Z39.48-1984.

Contents

Acknowledgments

Chemical Achievers is underwritten in part by educational grants from Glaxo Wellcome Inc. and the Shell Oil Company Foundation and represents the work of many hands over a number of years. The concept for this project was suggested by chemistry teachers participating in a summer workshop, sponsored by the Chemical Heritage Foundation and generously supported by the Dibner Fund, the Henkel Corporation, the Monsanto Company, and Union Carbide Corporation. Chemist-historians Derek Davenport and William Jensen taught this workshop and drew up the original list of chemical scientists relevant to the content of introductory chemistry courses—a list much revised over time.

Other historians who have rendered assistance along the way include Mi Gyung Kim, 1994 Edelstein Fellow, and Theodor Benfey, editor emeritus of *Chemical Heritage* and my most enduring and kindliest critic. Arthur Metzner and Mark Albers read the manuscript at an early stage and provided suggestions from the respective points of view of the chemical engineer and the high school teacher. Howard Sauertieg, research assistant in historical services, assembled the bibliography with enthusiasm and true diligence. Jean Hunsberger, now oral history assistant, lent word-processing assistance at critical moments.

Obtaining the visual images that are integral to giving "human faces" to the chemical science involved an even larger cast: Thelma McCarthy and Marjorie Gapp, successive curators of the CHF pictorial collections; research assistants Daniel Flaumenhaft and Liesl Allingham; and the many curators, archivists, and public relations specialists who supplied photographs. In this regard I owe a special debt to John Pollack, who cheerfully fulfilled our many requests for images from the Edgar Fahs Smith Collection, Special Collections, Van Pelt Library, University of Pennsylvania.

The chemists and chemical engineers whose accomplishments are profiled in this book and who are living history are to be thanked for their patience in answering my questions and their assistance in locating photographs of themselves.

Finally, the graceful presentation of text and images is due to the long labors of editors Patricia Wieland and Frances Kohler, and designer Sylvia Barkan.

Mary Ellen Bowden
Research Historian

Top to bottom: Alice Hamilton (Courtesy Michigan Historical Collections, Bentley Historical Library, University of Michigan); Charles F. Chandler (Courtesy Chandler Museum, Columbia University); Wallace Hume Carothers (Courtesy Hagley Museum and Library); Percy Lavon Julian (Courtesy DePauw University Archives and Special Collections); John Franz (Courtesy Monsanto Company).

The Human Face of Chemical Achievement

This book answers a request from teachers of courses in introductory chemistry at the high school and college levels for a source of pictorial material about famous chemical scientists that would be suitable as a teaching aid.

To answer the request, researchers at the Chemical Heritage Foundation collected biographical information and images on two types of achievers. First are the historical greats: those chemical scientists most often referred to in introductory courses. We have also included scientists who made contributions in areas of the chemical sciences that are of special relevance to modern life and the career choices students will make.

Teachers today want to present the human face of science, to point to the human beings who had the insights and made the major advances that they ask students to master. Boyle's law, for example, becomes somewhat less cold and abstract if we can connect it with a face, even if the face is topped with a wig. Marie Curie begins to be seen in the role of wife and mother as well as the role of genius scientist in photographs of her with her daughters—one of whom was also a Nobel Prize winner. And we are reminded of the ubiquity of the contribution of the chemical sciences to all aspects of our daily life when we read about Wallace Carothers's path to nylon, Percy Julian's work on hormones, and Charles Chandler's efforts to preserve the environment. Finally, this book also presents many images of chemists in their work setting—where chemistry is actually practiced.

The images, which come from the pictorial collections at CHF and at many other institutions, are presented in a format and special binding that allow for easy conversion to overhead transparencies. We welcome suggestions for improvements in our presentation and ways to make the book more useful, as well as for the names of chemical notables who might be made available in a second collection.

■ Robert Boyle at the age of thirty-seven, with his air pump in
the background. François Diodati reengraved this image from an
engraving by William Faithorne, *Opera varia* (1680). Courtesy
Edgar Fahs Smith Memorial Collection, Department of Special
Collections, University of Pennsylvania Library.

1. Forerunners

Three of the giants discussed in this section conform to the traditional picture of chemists as thinkers who pursue their understanding of the nature of matter in a laboratory setting. The fourth, however, points to another face of chemical achievement displayed in this work: that of the chemical industrialist.

ROBERT BOYLE (1627–1691)

Robert Boyle was born at Lismore Castle, Munster, Ireland, the fourteenth child of the Earl of Cork. As a young man of means he was tutored at home and on the Continent. He spent the later years of the English Civil Wars at Oxford, reading and experimenting with his assistants and colleagues. This group was committed to the New Philosophy, which valued observation and experiment at least as much as logical thinking in formulating accurate scientific understanding. At the time of the restoration of the British monarchy in 1660, Boyle played a key role in founding the Royal Society to nurture this new view of science.

Although Boyle's chief scientific interest was chemistry, his first published scientific work, *New Experiments Physico-Mechanicall, Touching the Spring of the Air and Its Effects* (1660), concerned the physical nature of air, as displayed in a brilliant series of experiments in which he used an air pump to create a vacuum. The second edition of this work, published in 1662, delineated the quantitative relationship that Boyle derived from experimental values, later known as "Boyle's law": that the volume of a gas varies inversely with pressure.

Boyle was an advocate of corpuscularism, a form of atomism that was slowly displacing Aristotelian and Paracelsian views of the world. Instead of defining physical reality and analyzing change in terms of Aristotelian substance and form and the classical four elements of earth, air, fire, and water—or the three Paracelsian elements of salt, sulfur, and mercury—corpuscularism discussed reality and change in terms of particles and their motion. Boyle believed that chemical experiments could demonstrate the truth of the corpuscularian philosophy. In this context he defined the term *element* in *Sceptical Chymist* (1661) as "certain primitive and simple, or perfectly unmingled bodies; which not being made of any other bodies, or of one another, are the ingredients of which all those called perfectly mixt bodies are immediately compounded, and into which they are

The Gift of the Right Hon:^{ble} the Earl of Burlington

■ Robert Boyle at the age of sixty-three, by James Worsdale after a painting by Johann Kerseboom. Boyle bequeathed funds to this American college to educate Native Americans to become Christian missionaries to their people. Courtesy Muscarelle Museum of Art, College of William and Mary.

ultimately resolved." He was probably referring to the *uniform* corpuscles—which were as yet unobserved—out of which corpuscular aggregates were formed, not using "elements" as Antoine-Laurent Lavoisier and others used the term in the eighteenth century to refer to *different* substances that could not be broken down further by chemical methods. In his experiments Boyle made many important observations, including that of the weight gain by metals when they are heated to become calxes. He interpreted this phenomenon as caused by fiery particles that were able to pass through the walls of glass vessels.

Boyle also wrote extensively on natural theology, advocating the notion that God created the universe according to definite laws.

JOSEPH PRIESTLEY (1733–1804)

Joseph Priestley, best remembered for his discovery of oxygen, was ceremoniously welcomed to the United States in 1794 as a leading contemporary thinker and friend of the new republic. Then sixty-one, he was known to Americans at least as well for his prodigious political and theological writings as for his scientific contributions.

Priestley was educated to be a minister in the churches that dissented from the Church of England, and he spent most of his life employed as a preacher or teacher. He gradually came to question the divinity of Jesus, while accepting much else of Christianity—in the process becoming an early Unitarian.

Priestley was a supporter of both the American and French Revolutions. He saw the latter as the beginning of the destruction of all earthly regimes that would precede the Kingdom of God, as foretold in the Bible. These freely expressed views were considered seditious by English authorities and many citizens. In 1791 a mob destroyed his house and laboratory in Birmingham. This episode and subsequent troubles made him decide to emigrate to the United States. With his sons he planned to set up a model community on undeveloped land in Pennsylvania, but like many such dreams, this one did not materialize. He and his wife did, however, build a beautiful home equipped with a laboratory far up the Susquehanna River in Northumberland.

Priestley's first scientific work, *The History of Electricity* (1767), was encouraged by Benjamin Franklin, whom he had met in London. In preparing the publication Priestley began to perform experiments—at first merely to reproduce those reported in the literature but later to answer questions of his own. In the 1770s he began his most famous scientific research on the nature and properties of gases. At that time he was living next to a brewery, which provided him an ample supply of carbon dioxide. His first *chemical* publication was a description of how to carbonate water, in imitation of some naturally occurring bubbly mineral waters. Inspired by

Stephen Hales's *Vegetable Staticks* (first edition, 1727), which described the pneumatic trough for gathering gases over water, Priestley began examining *all* the "airs" that might be released from different substances. Many, following Aristotle's teachings, still believed there was only one "air." By clever design of apparatus and careful manipulation, Priestley isolated and characterized eight gases, including oxygen—a record not equaled before or since. In addition, he contributed to the understanding of photosynthesis and respiration.

Priestley fought a long-running battle with Antoine-Laurent Lavoisier and his followers over how to interpret the results of experiments with gases. He interpreted them in terms of phlogiston—the hypothetical principle of flammability that was thought to give

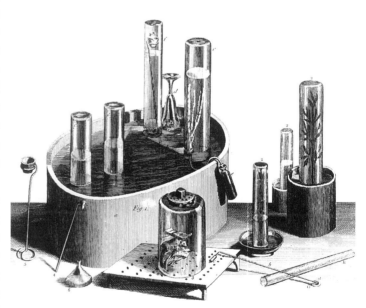

■ Priestley's pneumatic trough and other chemical apparatus. Note the mice in the container in the foreground used to test respirability of various gases. From *Experiments and Observations on Different Kinds of Air* (1774). Courtesy The Chemists' Club.

metals their luster and ductility and was widely used in the early eighteenth century to explain combustion, calcination, smelting, respiration, and other chemical processes. Proponents of phlogiston did not consider it to be a material substance, so it was therefore unweighable. Priestley gave qualitative explanations of these phenomena, talking, for example, about oxygen as "dephlogisticated air."

■ Joseph Priestley calling for the head of King George III on a platter, although he did not actually attend the Bastille Day celebration. Even so, his house and laboratory were destroyed that evening by a "Church and King" mob. Caricature by James Gillray. Gift of Derek Davenport.

■ Joseph Priestley, "Doctor Phlogiston," explaining away the Bible and expounding other incendiary views. Gift of Margaret B. Aydelotte.

■ Joseph Priestley, when he was minister at Leeds, where he conducted some investigations of oxygen and other gases. Gift of Margaret B. Aydelotte.

Antoine-Laurent Lavoisier (1743–1794)

The son of a wealthy Parisian lawyer, Antoine-Laurent Lavoisier completed a law degree in accordance with family wishes. His real interest, however, was in science, which he pursued with passion while leading a full public life. On the basis of his earliest scientific work, mostly in geology, he was elected in 1768—at the early age of twenty-five—to the Academy of Sciences, France's most elite scientific society. In the same year he bought into the Ferme Générale, the private corporation that collected taxes for the Crown on a profit-and-loss basis. A few years later he married the daughter of another tax farmer, Marie-Anne Pierrette Paulze, who was not quite fourteen at the time. Madame Lavoisier prepared herself to be her husband's scientific collaborator by learning English to translate the work of British chemists like Priestley and by studying art and engraving to illustrate Antoine-Laurent's scientific experiments.

In 1775 Lavoisier was appointed a commissioner of the Royal Gunpowder and Saltpeter Administration and took up residence in the Paris Arsenal. There he equipped a fine laboratory, which attracted young chemists from all over Europe to learn about the "Chemical Revolution" then in progress. He meanwhile succeeded in producing more and better gunpowder by increasing the supply and ensuring the purity of the constituents—saltpeter (sodium nitrate), sulfur, and charcoal—as well as by improving the methods of granulating the powder.

Characteristic of Lavoisier's chemistry was his systematic determination of the weights of reagents and products involved in chemical reactions, including the gaseous components, and his underlying belief that matter—identified by weight—would be conserved through any reaction. Among his contributions to chemistry associated with this method were the understanding of combustion and respiration as caused by chemical reactions with the part of the air he called "oxygen," and his definitive proof by composition and decomposition that water is made up of oxygen and hydrogen. His giving new names to substances—most of which are still used today—was an important means of forwarding the Chemical Revolution, because these terms expressed the theory behind them. In the case of "oxygen," from the Greek meaning "acid-former," Lavoisier expressed his theory that oxygen was the acidifying principle. He considered thirty-three substances as "elements"—by his definition, substances that chemical analyses had failed to break down into simpler entities. Ironically, considering his opposition to phlogiston, among these substances was "caloric"—the unweighable substance of heat, and possibly light, that caused other substances to expand when it was added to them. To propagate his ideas, in 1789 he published a textbook, *Traité élémentaire de chimie*, and began a journal, *Annales de Chimie*, which carried research reports about the new chemistry almost exclusively.

A political and social liberal, Lavoisier took an active part in the events leading to the French Revolution, and in its early years he drew up plans and reports advocating many reforms, including the establishment of the metric system of weights and measures. Despite his eminence and his services to science and France, he came under attack as a former farmer-general of taxes and was guillotined in 1794. A noted mathematician, Joseph-Louis Lagrange, remarked of this event, "It took them only an instant to cut off that head, and a hundred years may not produce another like it."

■ Antoine-Laurent Lavoisier conducts an experiment on human respiration in this drawing made by his wife, who depicted herself at the table on the far right. Courtesy Edgar Fahs Smith Memorial Collection, Department of Special Collections, University of Pennsylvania Library.

■ Antoine-Laurent Lavoisier and his wife, Marie Anne Pierrette Paulze, painted by Jacques Louis David, one of Marie's art instructors. Courtesy Metropolitan Museum of Art, Purchase, Mr. and Mrs. Charles Wrightsman Gift, 1977.

ELEUTHÈRE IRÉNÉE DU PONT (1771–1834)

Among the young men whom Lavoisier deeply influenced was Eleuthère Irénée du Pont, the founder of the DuPont Company. His father, Pierre Samuel du Pont—an economist, government official, and publicist—was among those attempting moderate reforms to inequitable and inefficient institutions in the last years of the Old Regime and the early days of the French Revolution. Pierre and Lavoisier became friends because of their similar political agendas, and together they reorganized the Royal Gunpowder and Saltpeter Administration. Lavoisier, one of the agency's four directors, soon hired Pierre's younger son, Eleuthère Irénée, to work in the Essonne gunpowder factory, where he learned how to manufacture gunpowder.

During the French Revolution the du Ponts found themselves under attack, not least because they had personally protected the king and queen from a mob besieging the Tuileries Palace in Paris in 1792. Only changes in administration saved Pierre Samuel from the guillotine. In 1799 he and his entire family left for the United States, where they hoped to found a model community of French exiles—which never came to pass.

Eleuthère Irénée soon recognized a business opportunity in the poor quality of the gunpowder generally available in the United States, and in 1802 he set up a powder works on the banks of the Brandywine River in Delaware. Start-up was difficult, and for the thirty-two years of his leadership he carried substantial debt, including loans to pension the widows and orphans of forty workers killed in a terrible explosion at the powder works in 1818. He nonetheless created a major American business enterprise, one that employed 140 men by 1827 and was producing over one million pounds of gunpowder per year by 1834, the year of his death.

■ First drawing of the du Pont powder mills, done in 1806 by Charles Dalmas, du Pont's brother-in-law and a workman in the mills. Houses and stores for the workers and their families are also shown. Courtesy Hagley Museum and Library.

■ Engraving of Eleuthère Irénée du Pont, after a painting by Rembrandt Peale. Courtesy Hagley Museum and Library.

2. Theory and Production of Gases

Lavoisier's Chemical Revolution was based on a theory that involved the chemical role of a gas—his oxygen theory—and his immediate and later successors further explored the character of gases. Their theoretical advances eventually proved of great importance to modern society: Many industrial processes require gases and their compounds and rely on a thorough understanding of the reactions that produce them.

JOSEPH LOUIS GAY-LUSSAC (1778–1850)

Joseph Louis Gay-Lussac grew up during both the French and Chemical Revolutions. His comfortable existence as the privately tutored son of a well-to-do lawyer was disrupted by political and social upheavals: His tutor fled, and his father was imprisoned. Joseph, however, benefited from the new order when he was selected to attend the Ecole Polytechnique, an institution of the French Revolution designed to create scientific and technical leadership, especially for the military. There his mentors included Pierre Simon de Laplace and Claude Louis Berthollet, among other scientists converted by Lavoisier to oxygen chemistry. Gay-Lussac's own career as a professor of physics and chemistry began at the Ecole Polytechnique.

He shared the interest of Lavoisier and others in the quantitative study of the properties of gases. From his first major program of research in 1801–1802, he concluded that equal volumes of all gases expand equally with the same increase in temperature: This conclusion is usually called "Charles's law" in honor of Jacques Charles, who had arrived at nearly the same conclusion fifteen years earlier but had not published it. In 1804 Gay-Lussac made several daring ascents of over seven thousand meters above sea level in hydrogen-filled balloons—a feat not equaled for another fifty years—that allowed him to investigate other aspects of gases. Not only did he gather magnetic measurements at various altitudes, but he also took pressure, temperature, and humidity measurements and samples of air, which he later analyzed chemically. In 1808 Gay-Lussac announced what was probably his single greatest achievement: From his own and others' experiments he deduced that gases at constant temperature and pressure combine in simple numerical proportions by volume, and the resulting product or products—if gases—also bear a simple proportion by volume to the volumes of the reactants.

With his fellow professor at the Ecole Polytechnique,

Louis Jacques Thénard, Gay-Lussac also participated in early electrochemical research, investigating the elements discovered by its means. Among other achievements, they decomposed boric acid by using fused potassium, thus discovering the element boron. The two also took part in contemporary debates that modified Lavoisier's definition of acids and furthered his program of analyzing organic compounds for their oxygen and hydrogen content.

■ Joseph-Louis Gay-Lussac and Jean-Baptiste Biot in their balloon on 24 August 1804. From Louis Figuier, *Les Merveilles de la Science* (4 vols.; Paris, 1867–70), *2*, p. 537.

■ Joseph Louis Gay-Lussac. Courtesy Edgar Fahs Smith Memorial Collection, Department of Special Collections, University of Pennsylvania Library.

ALFRED NOBEL (1833–1896)

The growing understanding of gases and the reactions that produce them was of great importance to modern industrial society. Not least was the production of explosives—substances that undergo reactions involving the release of heat and rapidly expanding gaseous products. In making black powder Lavoisier and E. I. du Pont were improving a technology known to Western cultures since the fourteenth century and even earlier in China and the Far East. In the nineteenth century much more powerful explosives were created by treating various organic substances with nitric acid. Among these new explosives was dynamite, a stabilized form of nitroglycerin, invented in 1867 by Alfred Nobel. One thousand times more powerful than black powder, it expedited the building of roads, tunnels, canals, and other construction projects worldwide.

Nobel's father, Immanuel, was a Swedish inventor-entrepreneur in St. Petersburg who supplied the Russian military with war matériel, including early underwater mines. Alfred and his brothers were educated at home by Swedish and Russian tutors in chemistry and other subjects. Alfred became very proficient in chemistry but also entertained ambitions of becoming a writer. Partly to dissuade him from the latter, his father financed his sixteen-year-old son's travel and study in Europe, including a stay of some months in the Paris laboratory of Théophile Pelouze, where Nobel shared workspace with an Italian chemist, Ascanio Sobrero, who had first prepared nitroglycerin in 1846. When Alfred was seventeen, he apprenticed in New York with the Swedish-American inventor John Ericsson, who later built the *Monitor*, the Union's ironclad warship that became famous for defeating the Confederacy's *Merrimac*. When Alfred returned to St. Petersburg, the Nobel factory was booming thanks to the Crimean War. When the war ended and the firm went into bankruptcy, Alfred and his father turned to developing methods to produce nitroglycerin in quantity. In 1862 Alfred began its manufacture in a small plant outside Stockholm—a venture that cost the life of his youngest brother, Emil. Alfred persevered, first inventing the blasting cap and then discovering that a silicaceous earth, kieselguhr, would stabilize nitroglycerin, thus making dynamite.

Alfred became wealthy by setting up companies and selling patent rights to dynamite and related products worldwide. The DuPont Company in the United States became one of the chief companies associated with Nobel. In 1875 Nobel created blasting gelatin, a colloidal suspension of nitrocellulose in glycerin, and in 1887 ballistite, a nearly smokeless powder especially suitable for propelling military projectiles. Nobel, the man who had tried to make handling explosives safe for workmen, was deeply troubled by the destructiveness of his inventions and became concerned with establishing worldwide peace.

Nobel died in 1896, leaving his considerable estate as an endowment for annual awards in chemistry, physics, medicine or physiology, literature, and peace—all of which represented his lifelong interests.

■ Alfred Nobel. Courtesy Edgar Fahs Smith Memorial Collection, Department of Special Collections, University of Pennsylvania Library.

■ A stick of dynamite bearing the Nobel company's insignia. Courtesy Nitro Nobel, AB.

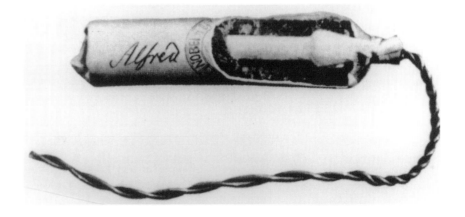

■ Carl von Linde. Courtesy Edgar Fahs Smith Memorial Collection, Department of Special Collections, University of Pennsylvania Library.

Carl von Linde (1842–1934)

Although discovering oxygen and investigating its role in chemical reactions proved to be of crucial importance in changing the science of chemistry, initially oxygen could be produced only in the laboratory and in limited quantities, by chemical or electrolytic means: It had little importance outside the laboratory. It was the achievement of Carl von Linde in 1902 to take oxygen from the air itself—and he was soon extracting it in quantities approaching one thousand cubic feet per hour. Oxygen became a common commodity that was supplied to hospitals and industries and was later used in rocket fuel, but this was not the German engineer's first important contribution.

Linde, the son of a Lutheran minister, was educated in science and engineering at the Federal Polytechnic in Zurich, Switzerland. After working for locomotive manufacturers in Berlin and Munich, he became a faculty member at the Polytechnic in Munich. His research there on heat theory, from 1873 to 1877, led to his invention of the first reliable and efficient compressed-ammonia refrigerator. The company he

established to promote this invention was an international success: Refrigeration rapidly displaced ice in food handling and was introduced into many industrial processes.

After a decade Linde withdrew from managerial activities to refocus on research, and in 1895 he succeeded in liquefying air by first compressing it and then letting it expand rapidly, thereby cooling it. He then obtained oxygen and nitrogen from the liquid air by slow warming. In the early days of oxygen production the biggest use by far for the gas was the oxyacetylene torch, invented in France in 1904, which revolutionized metal cutting and welding in the construction of ships, skyscrapers, and other iron and steel structures.

One company formed to utilize Linde's later patents was the Linde Air Products Company, founded in Cleveland in 1907. In 1917 Linde Air Products joined with four other companies that produced acetylene among other products to form Union Carbide and Carbon Corporation. Recently, Linde Air again became an independent company—Praxair.

■ Cylinders of oxygen being loaded on a tractor-trailer truck (1914) owned by the Linde Air Products Company. Courtesy Praxair, Inc.

FRITZ HABER (1868–1934)

By 1905 Fritz Haber had reached the objective long sought by chemists of *fixing* nitrogen from air. Using high pressure and a catalyst, he directly reacted nitrogen gas, which was generated by the Linde process, and hydrogen gas to create ammonia. The process was soon scaled up by BASF's great chemist and engineer Carl Bosch—hence the name "Haber-Bosch" process. The nitric acid produced from the ammonia was then used to manufacture agricultural fertilizers as well as explosives.

Haber was from a well-to-do German-Jewish family involved in various manufacturing enterprises. He studied at several German universities, earning a doctorate in organic chemistry in 1891. After a few years of moving from job to job, he settled into the Department of Chemical and Fuel Technology at the Polytechnic in Karlsruhe, Germany, where he mastered the new subject of physical chemistry. His research in physical chemistry eventually led to the Haber-Bosch process. In 1911 he was invited to become director of the Institute for Physical Chemistry and Electrochemistry at the new Kaiser Wilhelm Gesellschaft in Berlin, where academic scientists, government, and industry cooperated to promote original research.

The Haber-Bosch process is generally credited with keeping Germany supplied with fertilizers and munitions during World War I, after the British naval blockade cut off supplies of nitrates from Chile. During the war Haber threw his energies and those of his institute into further support for the German side. He developed a new weapon—poison gas, the first example of which was chlorine gas—and supervised its initial

■ The laboratory apparatus designed by Fritz Haber and Robert Le Rossignol for producing ammonia from hydrogen and nitrogen, which was scaled up in the Haber-Bosch process. The catalytic process took place in the large cylinder at the left. Courtesy Archiv zur Geschichte der Max-Planck-Gesellschaft, Berlin-Dahlem.

■ Fritz Haber, sketched in 1911 by W. Luntz.

deployment on the Western Front at Ypres, Belgium, in 1915. His promotion of this frightening weapon precipitated the suicide of his wife, who was herself a chemist, and many others condemned him for his war-time role. There was great consternation when he was awarded the Nobel Prize in chemistry for 1918 for the synthesis of ammonia from its elements.

After World War I, Haber was remarkably successful in building up his institute, but in 1933 the anti-Jewish decrees of the Nazi regime made his position untenable. He retired a broken man, although at the time of his death he was on his way to investigate a possible senior research position at Rehovot in Palestine (now Israel).

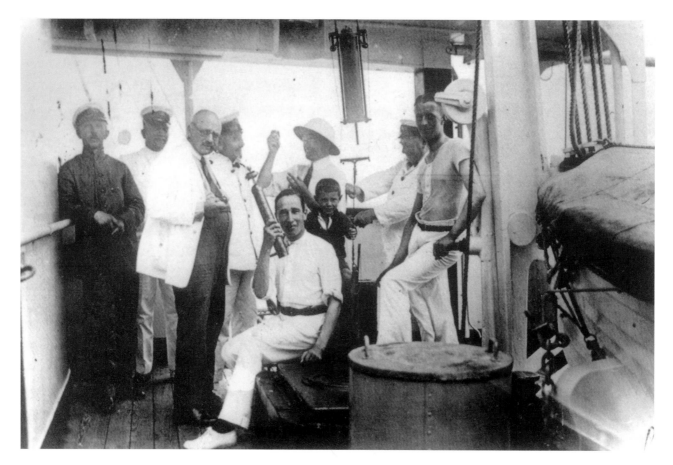

■ Fritz Haber, third from left, on board ship to Buenos Aires. He hoped to mine the ocean's minuscule percentage of gold to pay Germany's reparations imposed by the Versailles treaty that signaled the end of World War I. Courtesy Archiv zur Geschichte der Max-Planck-Gesellschaft, Berlin-Dahlem.

3. Electrochemistry and Electrochemical Industries

Electrochemistry has many applications beyond the university laboratory and provided the start of many modern corporations. These twelve achievers all investigated the connection between chemistry and electricity. The field began as a theoretical exploration of fundamental forces at work in the universe; it is now a cornerstone of industry, technology, and medicine.

HUMPHRY DAVY (1778–1829)

Humphry Davy, son of an impoverished Cornish woodcarver, rose meteorically to become a leader in the reformed chemistry movement initiated by Lavoisier—albeit a critic of some of its basic premises—and a pioneer in the new field of electrochemistry.

Apprenticed to an apothecary-surgeon, Davy taught himself a wide range of other subjects: theology and philosophy, poetics, seven languages, and several sciences, including chemistry. In 1798 he took a position at Thomas Beddoes's Pneumatic Institution, where the use of the newly discovered gases in the cure and prevention of disease was investigated. Davy's earliest published work ("An Essay on Heat, Light, and the Combinations of Light," in *Contributions to Physical and Medical Knowledge, Principally from the West of England*, ed. Beddoes, 1799) was a refutation of Lavoisier's caloric, arguing, among other points, that heat is motion but light is matter. But his early reputation was made by his book *Researches, Chemical and Philosophical, chiefly con-* *cerning Nitrous Oxide . . . and its Respiration* (1799). His recommendation that nitrous oxide (laughing gas) be employed as an anesthetic in minor surgical operations was ignored, but breathing it became the highlight of contemporary social gatherings. In 1801 Davy was appointed—first as lecturer, then as professor of chemistry—to the Royal Institution in London, which he molded into a center for advanced research and for polished demonstration lectures delivered to audiences largely made up of fashionable gentlemen and ladies.

Soon after the Italian physicist Alessandro Volta announced the electric pile—an early type of battery—in 1800, Davy rushed into this new field and correctly realized that the production of electricity depended on a chemical reaction taking place. His electrochemical experiments led him to propose that the tendency of one substance to react preferentially with other substances—its "affinity"—is electrical in nature.

Among his many accomplishments, Davy discovered

■ Humphry Davy, as painted by Sir Thomas Lawrence,
next to one of his inventions, the miner's safety lamp.

Scientific Researches! — New Discoveries in PNEUMATICKS! — or — an Experimental Lecture on the Powers of Air.

■ A young Humphry Davy gleefully works the bellows in this caricature by James Gillray of experiments with laughing gas at the Royal Institution. The lecturer is Thomas Garrett, Davy's predecessor as professor of chemistry. Benjamin Thompson, Count Rumford, the founder of the Royal Institution, stands at the doorway.

several new elements. In 1807 he electrolyzed slightly damp fused potash and then soda—substances that had previously resisted decomposition and hence were thought by some to be elements—and isolated potassium and sodium. He went on to analyze the alkaline earths, isolating magnesium, calcium, strontium, and barium. Davy's recognition that the alkalis and alkaline earths were all *oxides* challenged Lavoisier's theory that oxygen was the principle of acidity. Later, Davy determined that not all acids contain oxygen—including muriatic acid (our hydrochloric acid), which, as Davy discovered, was not "oxymuriatic acid," as Lavoisier

thought. It contained only hydrogen and one other element—chlorine.

In the course of his career Davy was involved in many practical projects. For example, he wrote the first text on the application of chemistry to agriculture and designed a miner's lamp that surrounded the lamp's flame with wire gauze to dissipate its heat and thus inhibit ignition of the methane gas commonly found in mines.

Davy became a fellow of the Royal Society in 1803 and served as its president from 1820 to 1827. He was knighted in 1812 and created a baronet in 1818—two honors, among many, that he much enjoyed.

■ Jöns Jakob Berzelius in 1843, after he had been Secretary of the Royal Swedish Academy of Sciences for twenty-five years. Painted by O. J. Soedermark. Courtesy Royal Swedish Academy of Sciences.

Jöns Jakob Berzelius (1779–1848)

Jöns Jakob Berzelius was one of Humphry Davy's contemporaries and rivals. Like Davy he became an accomplished experimenter in the field of electrochemistry, but Berzelius's mind was much more systematic than Davy's: He was given to running programs of hundreds of experiments and then deriving organized generalizations from them.

Berzelius was born into a well-educated Swedish family, but he experienced a difficult childhood because first his father and then his mother died. While in medical school at the University of Uppsala, he read about Volta's "pile" and immediately constructed one for himself. His thesis for his medical degree was on the effect of electric shock on patients with various diseases. Even though he reported no improvement in his patients, his interest in electrochemical topics continued. In 1807 he was made a professor at the Medical College in Stockholm—which soon after became the Karolinsska Institute—and a year later he began his long association with the Royal Swedish Academy of Sciences.

In preparing a chemistry textbook in Swedish for his medical students (*Lärboki Kemien,* vol. 1, 1808), Berzelius began the series of experiments for which he became most famous—those that established definitively that the elements in *inorganic* substances are bound together in definite proportions by weight (the law of constant proportions). His interest in all sorts of compounds led to his discovery of a number of new elements, including cerium, selenium, and thorium. Students working in his laboratory also discovered lithium, vanadium, and several rare earths. Using his experimental results, he determined the atomic weights of nearly all the elements then known. Dealing with so many elements in so many compounds motivated his creation of a simple and logical system of symbols—H, O, C, Ca, Cl, and so forth—which is basically the same as the system we use today, except that the combining proportions of the atoms of elements in a compound were indicated as superscripts instead of our subscripts. Ber-

zelius also applied his organizing abilities to mineralogy, where he classified minerals by their chemical composition rather than by their crystalline type, as had previously been done.

The major intellectual synthesis of Berzelius's career was "dualism"—a line of thinking that could be traced back to the original electrochemical investigations both he and Davy had made. Because compounds were decomposed by an electrical current and released elements were formed at the poles in an electrolytic cell, he

■ The "hungry" Berzelius. Presumed to be a portrait of Berzelius at some time between 1800 and 1810. Courtesy Royal Swedish Academy of Sciences.

assumed that atoms were charged and chemical combination resulted from the mutual neutralization of opposite charges. Dualistic thinking worked quite well except in the emerging realm of organic chemistry.

Berzelius was also a great organizer of men and institutions. As the Permanent Secretary of the Royal Swedish Academy of Sciences in Stockholm (1818–1848), he revived what had become a moribund organization. He continued to write textbooks, which were widely translated, and in 1822 he began a series of annual reports on the status of chemistry in Europe, which were also made available in other languages.

■ Jöns Jakob Berzelius welcoming King Karl XIV Johan (riding in the carriage) to the Royal Swedish Academy of Sciences. Courtesy Royal Swedish Academy of Sciences.

■ Michael Faraday in his laboratory at the Royal Institution. From a painting by Harriet Moore.

Michael Faraday (1791–1867)

The son of a poor and very religious family, Michael Faraday received little formal education. He was apprenticed to a bookbindery in London, however, and read many of the books brought there for binding, including the "electricity" section of the *Encyclopedia Britannica* and Jane Marcet's *Conversations on Chemistry*. He was also among the young Londoners who pursued an interest in science by gathering to hear talks at the City Philosophical Society. One of the bookbinder's customers gave Faraday free tickets to lectures given by Sir Humphry Davy at the Royal Institution, and after attending, Faraday conceived the goal of working for the great scientist. On the basis of Faraday's carefully taken notes of Davy's lectures, he was hired by Davy in 1813. His first assignment was to accompany Sir Humphry and his wife on a tour of the Continent, during which he sometimes had to be a personal servant to Lady Davy.

Once back in England, Faraday developed as an analytical and practical chemist. As his chemical capabilities increased, he was given more responsibility. In 1825 he replaced the seriously ailing Davy in his duties directing the laboratory at the Royal Institution. In 1833 he was appointed to the Fullerian Professorship of Chemistry—a special research chair created for him. Among other achievements, Faraday liquefied various gases, including chlorine and carbon dioxide. His investigation of heating and illuminating oils led to his discovery of benzene and other hydrocarbons, and he experimented at length with various steel alloys and optical glasses.

But Faraday is most famous for his contributions to the understanding of electricity and electrochemistry. In this work he was driven by his belief in the uniformity of nature and the interconvertibility of various forces, which he conceived early on as *fields* of force. In

1821 he succeeded in producing mechanical motion by means of a permanent magnet and an electric current—an ancestor of the electric motor. Ten years later, in 1831, he converted magnetic force into electrical force, thus inventing the world's first electrical generator. In the course of proving that electricities produced by various means are identical, Faraday discovered the two laws of electrochemistry: The amount of chemical change or decomposition is exactly proportional to the quantity of electricity that passes in solution; and the amounts of different substances deposited or dissolved by the same quantity of electricity are proportional to their chemical equivalent weights. In 1833 he and the classicist William Whewell worked out a new nomenclature for electrochemical phenomena based on Greek words, which is more or less still in use today—"ion," "electrode," and so on. Faraday suffered a nervous breakdown in 1839 but eventually returned to his electromagnetic investigations—this time on the relationship between light and magnetism. Although Faraday was unable to express his theories in mathematical terms, his ideas formed the basis for the electromagnetic equations that James Clerk Maxwell developed in the 1850s and 1860s.

In contrast to Davy, Faraday was known throughout his life as a kind and humble person, unconcerned with honors and eager to practice his science to the best of his ability.

PUNCH, OR THE LONDON CHARIVARI.—July 21, 1855.

FARADAY GIVING HIS CARD TO FATHER THAMES;
And we hope the Dirty Fellow will consult the learned Professor.

■ Michael Faraday's concern about contemporary environmental problems caricatured. From *Punch,* 21 July 1855.

■ Michael Faraday with an early electrical battery, after a painting by Thomas Phillips. Courtesy Edgar Fahs Smith Memorial Collection, Department of Special Collections, University of Pennsylvania Library.

■ Svante August Arrhenius. Courtesy Edgar Fahs Smith Memorial Collection, Department of Special Collections, University of Pennsylvania Library.

SVANTE AUGUST ARRHENIUS (1859–1927)

Svante August Arrhenius, a founding father of physical chemistry, was trained in both chemistry and physics. He began at the University of Uppsala, but then petitioned to work at the Royal Swedish Academy of Sciences in Stockholm instead, because he found the chemistry professors at the university rigid and uninspiring. His doctoral dissertation, presented in 1883, described his experimental work on the electrical conductivity of dilute solutions; it also contained a speculative section that set out an early form of his theory that molecules of acids, bases, and salts dissociate into ions when these substances are dissolved in water—in contrast to the notion of Faraday and others that ions are produced only when the electrical current begins to flow. According to Arrhenius, acids were substances that contained hydrogen and yielded hydrogen ions in aqueous solution; bases contained the OH group and yielded hydroxide ions in aqueous solution.

Arrhenius's thesis was received coolly by the university authorities and nearly ruined his prospects for an academic career. At the time his theory seemed incredible to many because, among other reasons, a

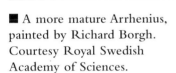

■ A more mature Arrhenius, painted by Richard Borgh. Courtesy Royal Swedish Academy of Sciences.

solution of sodium chloride shows none of the characteristics of either sodium or chlorine, and, in addition, the professors he had shunned in his studies were not well disposed toward him. But he had the foresight to send copies of his thesis to several international chemists, and a few were impressed with his work, including the young chemists Wilhelm Ostwald and Jacobus Henricus van't Hoff (see van't Hoff, Chapter 6), who were also to become founding fathers of physical chemistry. Ostwald offered Arrhenius a position in Riga, Latvia, which Arrhenius could not then accept because of his father's illness. He was instead given a post in Sweden and later a travel grant from the Swedish Academy that enabled him to work with Ostwald and van't Hoff. He subsequently developed his electrolytic dissociation theory further in quantitative terms and wrote texts promoting physical chemistry.

Arrhenius also applied physicochemical principles to the study of meteorology, cosmology, and biochemistry. In meteorology he anticipated late-twentieth-century speculation on the "greenhouse" effect of carbon dioxide in the atmosphere.

Although he was offered opportunities to move to other European universities, and he delivered important lecture series at universities in the United States, Arrhenius always returned to Stockholm. In 1903 he received the Nobel Prize in chemistry, and in 1905 he was made director of the newly created Nobel Institute for Physical Chemistry.

■ "Charged Croquet Balls." Drawing by William B. Jensen. Courtesy Oesper Collection, University of Cincinnati.

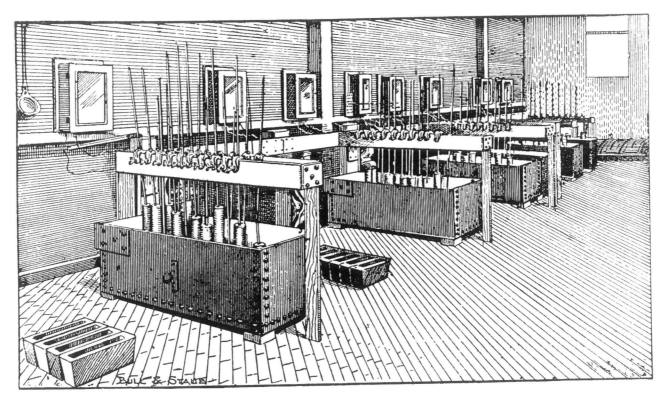

■ Drawing of the original Hall electrolytic cell set-up in the Pittsburgh Reduction Company plant, which shows the cast-iron crucibles or "pots"; the carbon anodes suspended by copper rods from an overhead copper support; and, on the floor, ingot molds. Courtesy Aluminum Company of America.

PAUL HÉROULT (1863–1914), CHARLES M. HALL (1863–1914), AND JULIA BRAINERD HALL (1859–1925)

Discovered in 1827 by Friedrich Wöhler (see Wöhler, Chapter 6), aluminum, though the most common metal on earth, is always found tightly locked in compounds. Efforts to use electrolysis to reduce it failed repeatedly, and for years it remained an exotic metal used in jewelry and for such special purposes as capping the Washington Monument.

The race for a commercially viable route to aluminum was won in 1886 by two young men working independently—Paul Héroult in France and Charles M. Hall in the United States. Hall was just six months out of Oberlin College; his sister Julia, who had also been a chemistry major at Oberlin, was of great assistance to him—helping with experiments, taking laboratory notes, and giving business advice.

The problem many researchers had with extracting aluminum was that electrolysis of an aluminum salt dissolved in water yields aluminum hydroxide. Both Hall and Héroult avoided this problem by dissolving aluminum oxide in a new solvent—fused cryolite, Na_3AlF_6. To scale up the process took Hall years of development and capital investment. In 1888 he joined with Alfred E. Hunt, an experienced metallurgist, to form the Pittsburgh Reduction Company. After exhausting the initial investment, the fledgling company was buoyed by the resources of the Mellon banking interests. Almost immediately the price of aluminum dropped dramatically. Developments in the early 1880s had reduced the price of a pound of aluminum from twelve dollars to four dollars a pound. The Hall process reduced it to two dollars a pound, and shortly after the company's move to Niagara Falls—the first electrochemical company in that location—to seventy-five cents and then thirty cents. In 1907 the company was renamed the Aluminum Company of America (Alcoa).

35

■ Paul Héroult. Courtesy Aluminum Company of America.

■ Charles M. Hall at the age of twenty-two. Courtesy Aluminum Company of America.

■ Julia Brainerd Hall, Charles's sister, who made many contributions to the discovery of the electrolytic means of producing aluminum and to the business success of the Pittsburgh Reduction Company and the Aluminum Company of America. Courtesy Aluminum Company of America.

■ Edward Goodrich Acheson in the lab with his omni-present cigar, testing Aquadag, a colloidal suspension of his artificial graphite. Courtesy Acheson Industries.

EDWARD GOODRICH ACHESON (1856–1931)

Edward G. Acheson was raised in the coal fields of south-western Pennsylvania. He left school at the age of sixteen to help support his family after his father died, but devoted his evenings to scientific pursuits—primarily electrical experiments. In 1880 he had the temerity to attempt to sell a battery of his own invention to Thomas Edison and wound up working for Edison at Menlo Park. After a year he was sent to Europe to install electrical lighting systems in the Hotel de Ville in Antwerp and La Scala in Milan, among other public places.

In 1884 Acheson left Edison's employ to become an independent inventor; he was soon successful. In 1891 he obtained the use of an electric generating plant of considerable power and tried to use electric heat to impregnate clay with carbon. The resultant mass exhibited some small shiny specks, and he determined that this crystalline substance had value as an abrasive—it was actually silicon carbide, which he called "carborundum." In 1894 he established the Carborundum Company in Monongahela City, Pennsylvania, to produce grinding wheels, whet stones, knife sharpeners, and powdered abrasives. In 1895 Acheson's electrochemical company was among the first to come to Niagara Falls. In its electric furnace he subsequently produced artificial graphite, another product that he commercialized, and he discovered that various organic substances allowed colloidal suspension of particles of graphite mixed in oil or water. His inventive genius knew no bounds; neither did his entrepreneurial optimism. Like many inventors, he was not a good manager, and his companies were constantly being taken out of his hands by

concerned investors. Many of Acheson's original companies live on today, including Carborundum, Inc., UCar International, and Acheson Industries.

■ Cover of the Carborundum Company's 1894 prospectus for its new diamond-like product. Courtesy the Carborundum Company.

HERBERT HENRY DOW (1866–1930)

Another early electrochemical pioneer was Herbert Henry Dow. The son of a master mechanic who earned his living by making improvements in small factories in Connecticut and Ohio, Dow possessed not only his father's inventive genius but even better business acumen. In a college project at Case School of Applied Science (now Case Western Reserve University), he analyzed samples of brine from the wells around Midland, Michigan, which contained small percentages of bromine in ionic form. In those days bromine had just a few uses: in medicines, "bromides," and photographic chemicals. The standard method of extracting bromine was to evaporate the brine, using the leftover wood scraps from the fast-disappearing lumber industry for fuel; remove the sodium chloride, which crystallized first; then add an oxidizing agent to the remaining liquid, which contained bromine ions; and finally distill the bromine.

To eliminate the need for the now-costly fuel used in the evaporation and distillation steps, Dow's ingenious plan, executed in 1889 in Canton, Ohio, was to oxidize the brine first with the proper amount of bleaching powder (calcium hydroxide, calcium chloride, and calcium hypochlorite), thus forming bromine, though it was still dissolved in brine. Next the brine was dripped onto burlap sacks, and in a "blowing out" process a current of air was passed through the brine-soaked sacks to carry off the bromine gas. The bromine-laden air was then brought into contact with iron or an alkali solution, and the bromine was thereby extracted from the air as $FeBr_2$, or alkali bromines. After Dow's first company went bankrupt, he moved to Midland in 1890 and organized the Midland Chemical Company. There he oxidized the brine by electrolysis, using electricity supplied by a second-hand 15-volt generator turned by the old steam engine in the flour mill he had rented in Midland.

After the bromine process was producing adequately, Dow next wanted to use electrolysis to make sodium hydroxide and chlorine to be turned into bleaching powder, but his first financial backers balked at this diversion of his talents. Dow turned instead to the faculty at Case for capital, forming the Dow Process Company in 1895. After considerable struggle, and yet another business reorganization to form the Dow Chemical Company, he again succeeded with a new process— the first of many successful diversifications and several business reorganizations, first into chlorine chemicals, then into organic chemicals, such as phenol and indigo dye, and finally into magnesium metal. During Dow's lifetime the company obtained its bromine, chlorine, sodium, calcium, and magnesium from the brine of ancient seas under Midland, but Dow, like Fritz Haber, developed experimental processes to mine modern seas. Three years after his death in 1930 his company opened its first seawater-processing plant in North Carolina. By World War II, Dow plants on the Gulf Coast were in a position to supply magnesium for firebombs and to make lightweight parts for airplanes.

■ Herbert H. Dow at the age of twenty-two. Courtesy Post Street Archives.

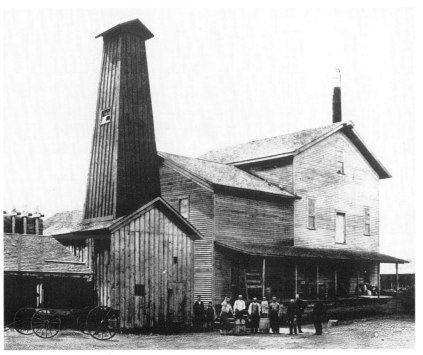

■ The old mill in which Dow set up his laboratory and his bromine process when he came to Midland in 1890. Courtesy Post Street Archives.

■ Søren Sørensen. Courtesy Oesper Collection, University of Cincinnati.

SØREN SØRENSEN (1868–1939)

Industry laboratories supplied theoretical concepts as well as industrial processes. The concept of pH was introduced in 1909 by the Danish chemist Søren Sørensen as a convenient way of expressing acidity—the negative logarithm of hydrogen ion concentration.

A Ph.D. from the University of Copenhagen, Sørensen was the director of the chemical department of the Carlsberg Laboratory, which was supported by the beer company of the same name—brewing being one of the oldest chemical industries. At the time, he was working on the effect of ion concentration in the analysis of proteins. Sørensen subsequently became a leader in the application of thermodynamics to proteins chemistry, and in this work he was assisted by his wife, Margrethe Høyrup Sørensen.

The context for the introduction of pH was the slow changeover from the old color-change tests for indicating the degree of acidity or basicity to electrical methods. In the latter, the current generated in an electrochemical cell by ions migrating to oppositely charged electrodes was measured, using a highly sensitive (and delicate) galvanometer. Until Sørensen developed the pH scale, there was no widely accepted way of expressing hydrogen ion concentrations.

■ Søren Sørensen visiting Cornell University in 1924. Courtesy Edgar Fahs Smith Memorial Collection, Department of Special Collections, University of Pennsylvania Library.

ARNOLD O. BECKMAN (1900–   )

Although close control of acidity is critical in the manufacture of many industrial products, industrial chemists continued to use color tests well into the twentieth century and rarely used the pH scale. The first commercially successful *electronic* pH meter was invented in 1934 by Arnold O. Beckman, then an instructor at the California Institute of Technology. A former classmate of his from the University of Illinois had the job of measuring the acidity of lemon juice for the California Fruit Growers' Association and asked Beckman to devise a sturdier electrical instrument for the task. To make his original pH meter sturdy, Beckman used the then recently invented vacuum tube. Although he was cautioned against starting up a company to offer a $195 instrument to scientists struggling to keep laboratories going in the middle of the Depression, Beckman went ahead, and the firm was a success. Among its other early products were an ultraviolet spectrophotometer—the Beckman DU (1940)—and an infrared and visible spectrophotometer—the Beckman IR-1 (1942). Today Beckman Instruments manufactures and markets instrument systems for conducting basic scientific research, new product research, and clinical diagnosis—and, of course, for students at all levels.

In 1940 Arnold Beckman gave up his faculty position at Caltech, but he remained deeply involved with education and research, serving on Caltech's board of trustees from 1953—as chairman from 1964 to 1974—and on the governing boards of several other colleges and universities. Through the Arnold and Mabel Beckman Foundation, the Beckmans have contributed substantially to the advancement of education and research nationwide.

■ Patent drawing for original Beckman pH meter (1936).

■ Arnold Beckman (right) working with his student
James McCullough at an optical bench in the 1930s.
Courtesy Archives, California Institute of Technology.

NORMAN BRUCE HANNAY (1921–1996)

Electrochemistry has also made a profound contribution to the age of computers, through the work of Norman Bruce Hannay. During World War II, Hannay worked on gaseous diffusion for the Manhattan Project, right after receiving his Ph.D. from Princeton in physical chemistry. His first civilian project at Bell Laboratories was to investigate the mechanism by which electrically charged particles are emitted from the incandescent cathodes of vacuum tubes—that is, along the same lines that Irving Langmuir at General Electric had pioneered (see Langmuir, Chapter 6). But at the end of 1947 Hannay's colleagues at Bell Labs invented the transistor, and his research program changed radically, to ensuring a high level of purity in the semiconductors used in transistors and later in integrated circuit chips.

Hannay's new job was to develop a mass spectrograph to analyze solids for trace impurities. Then he was chosen to lead both chemical and physical aspects of the silicon program. Although the first transistors were made from germanium, industry soon targeted silicon, with its lower cost and potential for developing a good oxide layer. To avoid contact with any other substance that would contribute impurities to silicon, Hannay's group devised a method of growing silicon crystals in a vacuum, relying on mere surface tension to suspend the crystals. This method is still used to make substrates for integrated circuit chips. In 1953 Bell Labs and Texas Instruments simultaneously heralded the arrival of the silicon transistor. Through the 1950s Hannay and his colleagues investigated other semiconductors, including gallium arsenide—the preferred material for semiconductor lasers, which are the basis for optical communication systems and have many other purposes today.

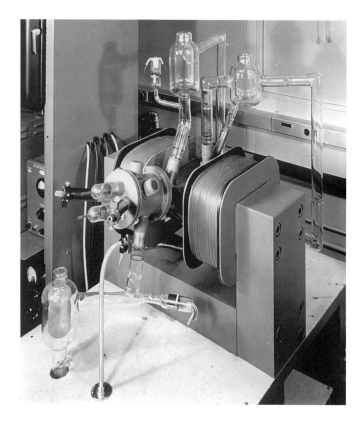

■ The mass spectrograph that Bruce Hannay developed for analyzing solids. Courtesy Bruce Hannay.

■ Bruce Hannay at Bell Laboratories in 1978. Courtesy AT&T Archives.

■ John Dalton, in an engraving after an 1814 painting by William Allen. Note the charts with Dalton's atomic symbols lying on the table.

4. The Path to the Periodic Table

The path to the periodic table begins early in the nineteenth century, when John Dalton united the atomic theory of matter—which had existed in various forms, including Boyle's notions, since antiquity—with the concept of the chemical element, which had emerged in the work of Lavoisier and his followers. On the basis of his newly synthesized theory, Dalton calculated the first relative weights of atoms and compounds. Although the method for calculating atomic weights was disputed for another fifty years, in the long run atomic weights would provide the key means of organizing the elements into the periodic table.

JOHN DALTON (1766–1844)

John Dalton was born into a modest Quaker family in Cumberland, England, and earned his living for most of his life as a teacher and public lecturer, beginning in his village school at the age of twelve. After teaching ten years at a Quaker boarding school in Kendal, he moved on to a teaching position in the burgeoning city of Manchester. There he joined the Manchester Literary and Philosophical Society, which provided him with a stimulating intellectual environment and laboratory facilities. The first paper he delivered before the society was on color blindness, which afflicted him and is sometimes still called "Daltonism."

Dalton arrived at his view of atomism by way of meteorology, in which he was seriously interested for a long period: He kept daily weather records from 1787 until his death, his first book was *Meteorological Observations* (1793), and he read a series of papers on meteorological topics before the Literary and Philosophical Society between 1799 and 1801. The papers contained Dalton's independent statement of Charles's law (see Gay-Lussac, Chapter 2): "All elastic fluids expand the same quantity by heat." He also clarified what he had pointed out in *Meteorological Observations*—that the air is not a vast chemical solvent as Lavoisier and his followers had thought, but a mechanical system, where the pressure exerted by each gas in a mixture is independent of the pressure exerted by the other gases, and where the total pressure is the sum of the pressures of each gas. In explaining the law of partial pressures to skeptical chemists of the day—including Humphry Davy—Dalton claimed that the forces of repulsion thought to cause pressure acted only between atoms of the same kind and that the atoms in a mixture were indeed different in weight and "complexity."

He proceeded to calculate atomic weights from percentage compositions of compounds, using an arbitrary system to determine the likely atomic structure of each compound. If there are two elements that can combine, their combinations will occur in a set sequence. The first compound will have one atom of A and one of B; the next, one atom of A and two atoms of B; the next, two atoms of A and one of B; and so on. Hence, water is HO. Dalton also came to believe that the particles in different gases had different volumes and surrounds of caloric, thus explaining why a mixture of gases—as in the atmosphere—would not simply layer out but was kept in constant motion. Dalton consolidated his theories in his *New System of Chemical Philosophy* (1808–1827).

As a Quaker, Dalton led a modest existence, although he received many honors later in life. In Manchester more than forty thousand people marched in his funeral procession.

■ Elements and their combinations as described in John Dalton's *New System of Chemical Philosophy* (1808–1827). Among the binary compounds, see number 21, which is water; 22, ammonia; 23, nitrous gas. Among the ternary, 26, nitrous oxide; 27, "nitric acid" (our NO_2).

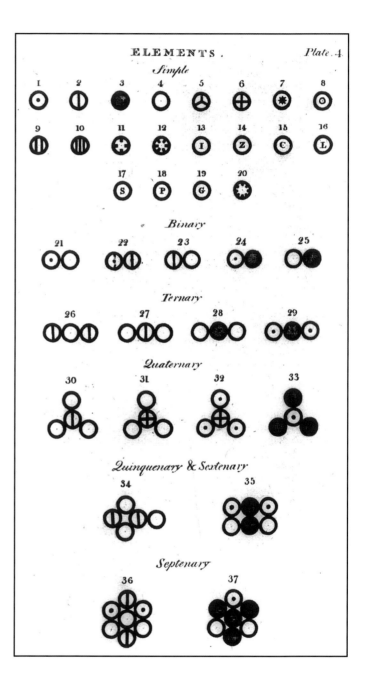

DALTON COLLECTING MARSH FIRE GAS

■ John Dalton collecting marsh gas (methane) with the aid of a boy assistant, as depicted in the Manchester Town Hall by Ford Madox Brown, who completed a series of murals on the city's history at the end of the nineteenth century. Courtesy Manchester City Council.

■ Caricature of John Dalton as President of the Literary and Philosophical Society of Manchester, ascribed to J. Derome and engraved by James Stephenson. Courtesy Edgar Fahs Smith Memorial Collection, Department of Special Collections, University of Pennsylvania Library.

■ Amedeo Avogadro. Courtesy Edgar Fahs Smith Memorial Collection, Department of Special Collections, University of Pennsylvania Library.

AMEDEO AVOGADRO (1776–1856)

Gay-Lussac's law of combining volumes (1808) (when two gases react, the volumes of the reactants and products—if gases—are in whole number ratios) tended to support Dalton's atomic theory (see Gay-Lussac, Chapter 2). Dalton did not in fact accept Gay-Lussac's work, but the Italian chemist Amedeo Avogadro saw it as the key to a better understanding of molecular constituency.

In 1811 Avogadro hypothesized that equal volumes of gases at the same temperature and pressure contain equal numbers of molecules. From this hypothesis it followed that relative molecular weights of any two gases are the same as the ratio of the densities of the two gases under the same conditions of temperature and pressure. Avogadro also astutely reasoned that simple gases were not formed of solitary atoms but were instead *compound* molecules of two or more atoms. (Avogadro did not actually use the word "atom"; at the time the words "atom" and "molecule" were used almost interchangeably. He talked about three kinds of "molecules," including an "elementary molecule"—what we would call an atom.) Thus Avogadro was able to overcome the difficulty that Dalton and others had encountered when Gay-Lussac reported that above 100°C the volume of water vapor was twice the volume of the oxygen used to form it. According to Avogadro, the molecule of oxygen had split into two atoms in the course of forming water vapor.

Curiously, Avogadro's hypothesis was neglected for half a century after it was first published. Many reasons for this neglect have been cited, including some theoretical problems, such as Berzelius's "dualism" (see Berzelius, Chapter 3), which asserted that compounds are held together by the attraction of positive and negative electrical charges, making it inconceivable that a molecule composed of two electrically similar atoms—as in oxygen—could exist. In addition, Avogadro was not part of an active community of chemists: The Italy of his day was far from the centers of chemistry in France, Germany, England, and Sweden, where Berzelius was based.

Avogadro was a native of Turin, where his father, Count Filippo Avogadro, was a lawyer and government leader in the Piedmont (Italy was then still divided into independent countries). Avogadro succeeded to his father's title, earned degrees in law, and began to practice as an ecclesiastical lawyer. After obtaining his formal degrees, he took private lessons in mathematics and sciences, including chemistry. For much of his career as a chemist he held the chair of physical chemistry at the University of Turin.

■ Stanislao Cannizzaro. Courtesy Edgar Fahs Smith Memorial Collection, Department of Special Collections, University of Pennsylvania Library.

STANISLAO CANNIZZARO (1826–1910)

In 1858, two years after Avogadro's death, his fellow Italian Stanislao Cannizzaro outlined a course in theoretical chemistry for students at the University of Genoa—where he had to teach without benefit of a laboratory. He used Avogadro's hypothesis as a pathway out of the confusion rampant among chemists about atomic weights and the fundamental structure of chemical compounds.

By all accounts Cannizzaro was much clearer in his explanations than Avogadro, and as an organic chemist he also showed how Avogadro's ideas could be applied to this branch of chemistry. In 1860 the first international chemical congress was held in Karlsruhe, Germany, to settle some of the contemporary chemical disputes—how to define "molecule" and "atom," what chemical nomenclature to use, how to determine atomic weights, and so on. After much discussion the chemists agreed to return home to decide for themselves how to proceed. However, many participants carried away a handout, a printed version of Cannizzaro's outline, that seemed convincing upon later reading.

At this time Cannizzaro was in the midst of eventful chemical and political careers. He was born in Palermo, Sicily, where his father was a magistrate and the minister of police, and he later attended medical school there, which kindled an interest in chemistry. Despite his family's connections to the royal court in Naples, he joined the antimonarchical 1848 revolution in Sicily. When it failed, he fled to Paris, where he resumed his chemical studies. After returning to Italy he held academic appointments in Alessandria, where he worked out the "Cannizzaro reaction"—the self-oxidation and self-reduction of aldehydes—and Genoa, where he expounded Avogadro's hypothesis. He next supported Giuseppe Garibaldi's Sicilian revolt of 1860 and took part in the new government centered in Palermo. During this time

he expanded the program of chemical studies at the university there. Upon Italian unification in 1871 he moved to Rome, where he continued his roles as a public figure and as a chemical scientist and educator.

■ Stanislao Cannizzaro at the age of thirty-two, after a sketch by Demetrio Salazzaro. **Courtesy Edgar Fahs Smith Memorial Collection, Department of Special Collections, University of Pennsylvania Library.**

■ Gustav Robert Kirchhoff (left) and Robert Wilhelm Eberhard Bunsen. Courtesy Edgar Fahs Smith Memorial Collection, Department of Special Collections, University of Pennsylvania Library.

ROBERT WILHELM EBERHARD BUNSEN (1811–1899) AND GUSTAV ROBERT KIRCHHOFF (1824–1887)

In 1860 Robert Bunsen and Gustav Kirchhoff discovered two alkali metals, cesium and rubidium, with the aid of the spectroscope they had invented the year before. These discoveries inaugurated a new era in the means used to find new elements. The first fifty elements discovered—beyond those known since ancient times—were either the products of chemical reactions or were released by electrolysis (see Chapter 3). From 1860 the search was on for *trace* elements detectable only with the help of specialized instruments like the spectroscope.

Bunsen, the son of a professor of modern languages at Göttingen University, earned his doctorate from that university in 1830. He was then given a three-year travel grant that took him to factories, places of geologic in-

terest, and famous laboratories, including Gay-Lussac's in Paris. Early in his career he did research in organic chemistry, which cost him the use of his right eye when an arsenic compound, cacodyl cyanide, exploded. Throughout his career he remained deeply interested in geological topics and once made daring temperature measurements of the water in the geyser tube of Iceland's Great Geyser just before it erupted.

Bunsen and Kirchhoff, a physicist trained at Königsberg, met and became friends in 1851, when Bunsen spent a year at the University of Breslau, where Kirchhoff was also teaching. Bunsen was called to the University of Heidelberg in 1852, and he soon arranged for Kirchhoff to teach at Heidelberg as well.

Bunsen's most important work was in developing

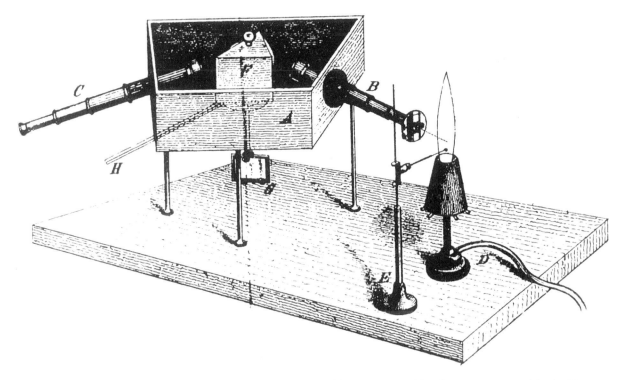

■ Bunsen-Kirchhoff spectroscope with the Bunsen burner (labeled D) from *Annalen der Physik* (1860).

several techniques used in separating, identifying, and measuring various chemical substances. He also made a number of improvements in chemical batteries for use in isolating quantities of pure metals—including one known as the Bunsen battery. He created the Bunsen burner for use in flame tests of various metals and salts: Its nonluminous flame did not interfere with the colored flame given off by the test material.

This line of work led to the spectroscope. It was Kirchhoff who suggested that similarly colored flames could possibly be differentiated by looking at their emission spectra through a prism. When he shone bright light through such flames, the dark lines in the absorption spectrum of the light corresponded in wavelengths, with the wavelengths of the bright, sharp lines characteristic of the emission spectra of the same test materials.

Bunsen spent the last forty years of his career at Heidelberg. Young chemists flocked to him, including Julius Lothar Meyer and Dmitri Mendeleev.

■ "Burner." Drawing by William B. Jensen. Courtesy Oesper Collection, University of Cincinnati.

JULIUS LOTHAR MEYER (1830–1895) AND DMITRI IVANOVICH MENDELEEV (1834–1907)

Julius Lothar Meyer and Dmitri Ivanovich Mendeleev worked with Bunsen at Heidelberg only five years apart, but they arrived there with significantly different backgrounds. Meyer was virtually born into a scientific career. He came from a medical family of Oldenburg, Germany, and first pursued a medical degree. In medical school he became interested in chemistry, especially physiological topics like gases in the blood. Mendeleev was born in Tobolsk, Siberia, where his father taught Russian literature and his mother owned and operated a glassworks. His early contacts with political exiles gave him a lifelong love of liberal causes, and his freedom to roam the glassworks stimulated an interest in business and industrial chemistry. His mother—after her husband's death and shortly before her own—took the fifteen-year-old Dmitri to St. Petersburg. There he attended the Main Pedagogical Institute and the University of St. Petersburg, where he pursued a doctorate in chemistry. During his graduate studies he traveled to Heidelberg to work with Bunsen.

Meyer and Mendeleev were among the young chemists attending the Karlsruhe Congress in 1860, and both were impressed with Cannizzaro's presentation of Avogadro's hypothesis (see earlier in this chapter). For both, writing a textbook proved to be the impetus for developing the periodic table—that is, a device to present the more than sixty known elements in an intelligible fashion. For some time chemists had been trying to devise a logical system of classification by arranging the elements by atomic weight, but confusion over how to determine atomic weights thwarted their attempts. Soon after Karlsruhe, various new atomic arrangements were published, culminating in the work of Meyer and Mendeleev. In the first edition of *Die Modernen Theorien der Chemie* (1864), Meyer used atomic weights to arrange twenty-eight elements into six families that bore similar chemical and physical characteristics, leaving a blank for an as-yet-undiscovered element. His one con-

ceptual advance over his immediate predecessors was seeing valence, the number that represents the combining power of an element (e.g., with atoms of hydrogen), as the link among members of each family of elements and as the pattern for the order in which the

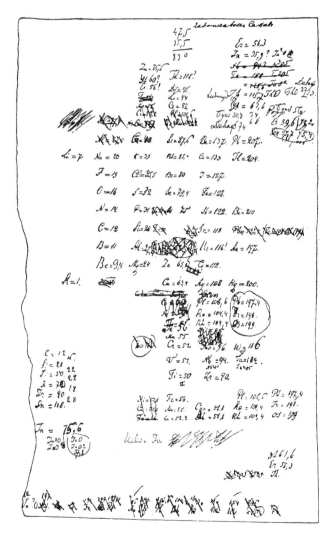

■ Draft for the first version of Mendeleev's periodic table (17 February 1869). Courtesy Oesper Collection, University of Cincinnati.

■ Julius Lothar Meyer. Courtesy Edgar Fahs Smith Memorial Collection, Department of Special Collections, University of Pennsylvania Library.

■ Dmitri Ivanovich Mendeleev in his study at home in 1904. Courtesy Edgar Fahs Smith Memorial Collection, Department of Special Collections, University of Pennsylvania Library.

families were themselves organized. In his original scheme the valences of the succeeding families, beginning with the carbon group, were 4, 3, 2, 1, 1, and 2.

After returning to St. Petersburg from Karlsruhe, Mendeleev taught at the St. Petersburg Technological Institute, completed his doctoral dissertation, started an experimental farm, and lectured for the Free Economic Society on agricultural topics. When in 1867 he was appointed to the chair of chemistry at the University of St. Petersburg, he too began to write a textbook, *Osnovy Khimii* (*Principles of Chemistry*; first edition, 1871), and

worked out the "periodic law," which was first published in papers in 1869. Mendeleev succeeded in arranging *all* known elements into one table.

Meyer then published his classic paper of 1870 ("Die Natur der chemischen Elemente als Function ihrer Atomgewichte," *Justus Liebigs Annalen der Chemie*, supp. 7 [1870], 354–364), describing the evolution of his work since 1864. This paper is particularly famous for its graphic display of the periodicity of atomic volume plotted against atomic weight. Many chemists, including Bunsen, had their doubts about the periodic law at first,

■ Dmitri Ivanonich Mendeleev as a young man. Courtesy Edgar Fahs Smith Memorial Collection, Department of Special Collections, University of Pennsylvania Library.

but these doubters were gradually converted by the discovery of elements predicted by the tabular arrangement and the correction of old atomic weights that the table cast in doubt. Meanwhile, Meyer and Mendeleev carried on a long drawn-out priority dispute.

Whereas Meyer continued to pursue a life of research and teaching and spent the last twenty years of his life as a professor at Tübingen, Mendeleev's strong democratic leanings got him into trouble with political and academic authorities, although his scientific eminence and the usefulness of his advice protected him to a certain degree. But in 1890 he left his professorship at the University of St. Petersburg after an official rebuke for delivering a student protest to the ministry of education. He then rose in government service to the position of Director of the Central Board of Weights and Measures. He contributed to the modernization of Russia through his reports and recommendations on weights and measures, protective tariffs, shipbuilding and shipping routes in Arctic regions, the manufacture of smokeless powder, and the development of heavy industry. When he died, students carried the periodic table in the funeral procession.

■ Periodic table on end wall of Institute of Weights and Measures in St. Petersburg. Courtesy Daniel Conlon.

■ William Ramsay. Courtesy Edgar Fahs Smith Memorial Collection, Department of Special Collections, University of Pennsylvania Library.

WILLIAM RAMSAY (1852–1916)

William Ramsay is known for work that established a whole new group in the periodic table—variously called over time inert, rare, or noble gases. In the last decade of the nineteenth century he and the famous physicist Lord Rayleigh (John William Strutt, 1842–1919)—already known for his work on sound, light, and other electromagnetic radiation—carried out separate investigations, for which they received Nobel Prizes in 1904, Ramsay in chemistry and Lord Rayleigh in physics.

Ramsay began his studies in his native city of Glasgow and completed a doctorate in chemistry at Tübingen, focusing on organic chemistry. On his return to Great Britain and his appointment to academic posts at the University of Bristol and then at University College, London, he became known for the inventiveness and scrupulousness of his experimental techniques, especially for his methods for determining the molecular weights of substances in the liquid state.

In 1892 Ramsay's curiosity was piqued by Lord Rayleigh's observation that the density of nitrogen extracted from the air was always greater than nitrogen released from various chemical compounds. Ramsay then set about looking for an unknown gas in air of greater density, which—when he found it—he named argon. While investigating for the presence of argon in a uranium-bearing mineral, he instead discovered helium, which since 1868 had been known to exist, but only in the sun. This second discovery led him to suggest the existence of a new group of elements in the periodic table. He and his coworkers quickly isolated neon, krypton, and xenon from the earth's atmosphere.

The remarkable inertness of these elements resulted in their use for special purposes, for example, helium instead of highly flammable hydrogen for lighter-than-air craft and argon to conserve the filaments in light bulbs. Their inertness also contributed to the "octet rule" in the theory of chemical bonding (see Lewis, Chapter 6). But in 1933 Linus Pauling (see Pauling, Chapter 6) suggested that compounds of the noble gases should be possible. Indeed, in 1962 Neil Bartlett, work-

ing at the University of British Columbia and later at Princeton, prepared the first noble gas compound—xenon hexafluoroplatinate, $XePtF_6$. Compounds of most of the noble gases have now been found.

■ William Ramsay as the personification of chemistry in *Vanity Fair*. Caricature by Spy. Courtesy Edgar Fahs Smith Memorial Collection, Department of Special Collections, University of Pennsylvania Library.

THEODORE WILLIAM RICHARDS (1868–1928)

Theodore William Richards, the first American to be awarded the Nobel Prize in chemistry, received it for his accurate determinations of atomic weights—twenty-five in all, including those used to determine virtually all other atomic weights. His work, which he began publishing in 1887, corrected earlier studies done in the 1860s by Jean Servais Stas. Among other contributions, Richards provided the experimental verification of the isotope concept, showing that lead from different sources has different atomic weights.

Richards was educated at home by his mother, a Quaker author and poet, and his father, a noted painter of seascapes, until he went to Haverford College at the age of fourteen. He proceeded to Harvard, where he earned a doctorate in chemistry by the time he was twenty. He remained there as an important researcher and teacher, except for two sojourns in Europe—first on a prize fellowship and, much later, to learn about the latest developments in electrochemistry and thermodynamics to pass on to his students.

■ Theodore W. Richards. Courtesy Edgar Fahs Smith Memorial Collection, Department of Special Collections, University of Pennsylvania Library.

5. Atomic
and
Nuclear Structure

Chemistry and physics overlap at the level where investigations of the smallest particles of matter are carried out. Appropriately, several of the achievers in this chapter are more commonly identified as physicists, but the line between a Seaborg and a McMillan, for example, is hard to draw, and the Nobel Prizes for this type of work were granted in both categories.

JOSEPH JOHN THOMSON (1856–1940)

In 1897 the physicist Joseph John (J. J.) Thomson discovered the electron in a series of experiments designed to study the nature of electric discharge in a high-vacuum cathode-ray tube—an area being investigated by numerous scientists at the time. Thomson interpreted the deflection of the rays by electrically charged plates and magnets as evidence of "bodies much smaller than atoms" that he calculated as having a very large value for the charge to mass ratio. Later he estimated the value of the charge itself. In 1904 he suggested a model of the atom as a sphere of positive matter in which electrons are positioned by electrostatic forces. His efforts to estimate the number of electrons in an atom from measurements of the scattering of light, X, beta, and gamma rays initiated the research trajectory along which his student Ernest Rutherford moved (see Rutherford, later in this chapter). Thomson's last important experimental program focused on determining the nature of positively charged particles. Here his techniques led to the development of the mass spectroscope, an instrument perfected by his assistant, Francis Aston, for which Aston received the Nobel Prize in 1922.

Ironically, Thomson—great scientist and physics mentor—became a physicist by default. His father intended him to be an engineer, which in those days required an apprenticeship, but his family could not raise the necessary fee. Instead young Thomson attended Owens College, Manchester, which had an excellent science faculty. He was then recommended to Trinity College, Cambridge, where he became a mathematical physicist. In 1884 he was named to the prestigious Cavendish Professorship of Experimental Physics at Cambridge, although he had personally done very little experimental work. Even though he was clumsy with his hands, he had a genius for designing apparatus and diagnosing its problems. He was a good lecturer, encouraged his students, and devoted considerable attention to the wider problems of science teaching at

■ J. J. Thomson in earlier days.

university and secondary levels. Of all the physicists associated with determining the structure of the atom, Thomson remained most closely aligned to the chemical community because his non-mathematical atomic theory—unlike early quantum theory—could also be used to account for chemical bonding and molecular structure (see G. N. Lewis and Irving Langmuir, Chapter 6).

Thomson received various honors, including the Nobel Prize in physics in 1906 and a knighthood in 1908. He also had the great pleasure of seeing several of his close associates receive their own Nobel Prizes.

■ J. J. Thomson (left) and Ernest Lord Rutherford in 1938.

ERNEST RUTHERFORD (1871–1937)

A consummate experimentalist, Ernest Rutherford was responsible for a remarkable series of discoveries in the fields of radioactivity and nuclear physics. He discovered alpha and beta rays, set forth the laws of radioactive decay, and identified alpha particles as helium nuclei. Most important, he postulated the *nuclear* structure of the atom: Experiments done in Rutherford's laboratory showed that when alpha particles are fired into gas atoms, a few are violently deflected, which implies a dense, positively charged central region containing most of the atomic mass.

Born on a farm in New Zealand, the second of twelve children, Rutherford completed a degree at the University of New Zealand and began teaching unruly schoolboys. He was released from this task by a scholarship to Cambridge University, where he became J. J. Thomson's (see earlier in this chapter) first graduate student at the Cavendish Laboratory. There he began experimenting with the transmission of radio waves, went on to join Thomson's ongoing investigation of the conduction of electricity through gases, and then turned to the field of radioactivity just opened up by Henri Becquerel and Pierre and Marie Curie (see later in this chapter).

Throughout his career Rutherford displayed his ability to work creatively with associates, some of whom were already established at the institutions to which he was appointed and others of whom he attracted as doctoral or postgraduate students. At McGill University in Montreal, his first appointment, he worked with Frederick Soddy on radioactive decay. At Manchester University he collaborated with Hans Geiger (of Geiger counter fame), Niels Bohr (whose model of atomic structure succeeded Rutherford's), and H. G. J. Moseley (who obtained experimental evidence for atomic numbers). During World War I, this Manchester research group was largely dispersed, and Rutherford turned to solving problems connected with submarine detection. After the war he succeeded J. J. Thomson in the Cavendish Professorship at Cambridge and again gathered a vigorous research group, including James Chadwick, the discoverer of the neutron.

Like his mentor Thomson, Rutherford garnered many honors. He received the Nobel Prize in physics for 1908; he was made a knight, then a peer with a seat in the House of Lords; and for the ultimate honor he was buried in Westminster Abbey.

■ Ernest Rutherford in academic garb. Courtesy Edgar Fahs Smith Memorial Collection, Department of Special Collections, University of Pennsylvania Library.

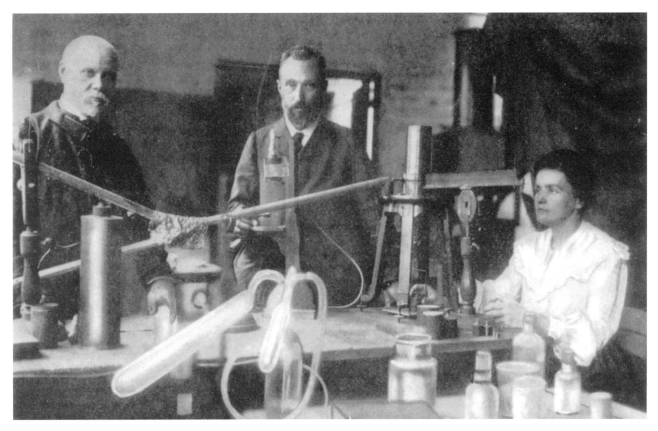

■ Pierre Curie, Pierre's assistant, Petit, and Marie Curie in the laboratory. Courtesy Archives Curie et Joliot-Curie, Paris.

MARIE SKLODOWSKA CURIE (1867–1934)

Marie Sklodowska Curie was the first person ever to receive two Nobel Prizes: the first in 1903 in physics, shared with her husband Pierre and Henri Becquerel for the discovery of the phenomenon of radioactivity; and the second in 1911 in chemistry for the discovery of the radioactive elements polonium and radium.

The daughter of impoverished Polish schoolteachers, Marie worked as a governess in Poland to support her older sister in Paris, whom she eventually joined. Already entranced with chemistry, Marie took advanced scientific degrees at the Sorbonne, where she met and married Pierre Curie, a physicist who had achieved fame for his work on the piezoelectric effect. For her thesis

she chose to work in a field just opened up by Wilhelm Roentgen's discovery of X rays and Becquerel's observation of the mysterious power of samples of uranium salts to expose photographic film. She soon convinced her husband to join in the endeavor of isolating the "radioactive" substance—a word she coined.

In 1898, after laboriously isolating various substances by successive chemical reactions and crystallizations of the products, which they then tested for their ability to ionize air, the Curies announced the discovery of polonium, and then of radium salts weighing about 0.1 gram that had been derived from tons of uranium ore. After Pierre's death in 1906 in a streetcar accident, Marie

■ *Right:* Marie Curie. Courtesy Edgar Fahs Smith Collection, Department of Special Collections, University of Pennsylvania Library. *Below:* Marie and her two daughters, Eve and Irène, in 1908. Courtesy Archives Curie et Joliot-Curie, Paris.

■ Marie and Pierre Curie caricatured in *Vanity Fair*, 22 December 1904.

achieved their objective of producing a pure specimen of radium.

Just before World War I, radium institutes were established for her in France and in Poland to pursue the scientific and medical uses of radioactivity. During the war Marie organized a field system of portable X-ray machines to help in treating wounded French soldiers.

In the midst of her busy scientific career Marie raised two daughters—in part, with the help of her father-in-law. Her elder daughter Irène became a Nobel Prize–winning chemist, also with her husband, Frédéric Joliot. Mother and daughter both eventually died of leukemia induced by their long exposure to radioactive materials.

■ Frédéric Joliot and Irène Joliot-Curie. Courtesy Edgar Fahs Smith Memorial Collection, Department of Special Collections, University of Pennsylvania Library.

IRÈNE JOLIOT-CURIE (1897–1956) AND FRÉDÉRIC JOLIOT (1900–1958)

Irène Joliot-Curie had the unusual experience of attending for two years in her childhood a special school that emphasized science, organized by her mother, Marie Curie, and her scientific friends for their own children. She was still a teenager when she worked with her mother in the radiography corps during World War I. After the war she assisted her mother at the Radium Institute in Paris, meanwhile completing her doctorate. She married Frédéric Joliot, a young physicist who had come to work with her mother.

The Joliot-Curies won the Nobel Prize for chemistry in 1935 for their discovery of artificial radiation by bombardment of alpha particles (helium nuclei, He^{2+}) on various light elements. They correctly interpreted the continued positron emission that occurred after bombardment had ceased as evidence that "radioactive isotopes" of known elements had been created. These isotopes rapidly became important tools in biomedical research and in the treatment of cancer.

The Joliot-Curies were the parents of a boy and a girl, both of whom became scientists—thus continuing a famous scientific dynasty.

OTTO HAHN (1879–1968), LISE MEITNER (1878–1968), AND FRITZ STRASSMANN (1902–1980)

In 1938 Otto Hahn, Lise Meitner, and Fritz Strassmann were the first to recognize that the uranium atom, when bombarded by neutrons, actually split.

With doctorate in hand from the University of Marburg in Germany, Hahn intended to make a career as an industrial chemist in a company with international business connections. He traveled to England to improve his English-language skills and found a job as an assistant in William Ramsay's laboratory at University College, London. Hahn quickly demonstrated his great skill as an experimentalist by isolating radioactive thorium. After working with Ernest Rutherford in Montreal, he joined Emil Fischer's institute at the University of Berlin, where he rose through the faculty ranks.

Hahn went in search of a collaborator with whom to pursue studies in experimental radioactivity and teamed up with Lise Meitner. She had come to Berlin to attend Max Planck's lectures in theoretical physics after receiving her doctorate in physics from the University of Vienna in 1905—the second doctorate in science from that university granted to a woman. In the first year of the Hahn–Meitner partnership they had to work in a remodeled carpenter's shop because the university did not yet accept women on an official basis. In 1912 their research group was relocated to the new Kaiser Wilhelm Gesellschaft, where Fritz Haber (see Haber, Chapter 2) was head of the physical chemistry institute, Hahn was head of the radioactivity institute, and from 1918, Meitner was head of the radioactivity institute's physics department. During World War I, Hahn served in the German gas warfare service headed by Haber, and Meitner volunteered as an X-ray nurse for the Austrian army.

The discovery of the neutron by James Chadwick in 1932 gave new impetus to radioactivity studies because this uncharged atomic particle could penetrate the secrets of the atomic nucleus more successfully. Meitner, Hahn, and another chemist, Fritz Strassmann, who had worked with the partners since 1929, were deeply involved in identifying the products of neutron bombardment of uranium and their decay patterns. It was generally expected that elements close in atomic number—quite possibly elements with higher atomic numbers than uranium—would be produced. In 1938 Meitner had to leave Berlin because the Nazis were closing in on all people of Jewish ancestry. She soon found a congenial setting for her research at the Nobel Institute in Stockholm. Her nephew, the physicist Otto Frisch, was located at Niels Bohr's institute in Copenhagen. Meanwhile, Hahn and Strassmann found that they had unexpectedly produced barium, a much lighter element than uranium, and they reported this news to Meitner. She and her nephew worked out the physics calculations of the phenomenon based on Bohr's "droplet" model of the nucleus and clearly stated that nuclear fission of uranium had occurred. It was quickly recognized that barium was among the stable isotopes that were the products of the radioactive decay of transuranic elements that must have been initially formed after neutron bombardment of uranium. News of the splitting of the atom and its awesome possibilities was brought by Bohr to scientists in the United States and ultimately resulted in the Manhattan Project.

Hahn, Meitner, and Strassmann were not engaged in nuclear weapons research during World War II. At the end of the war Hahn was astonished to hear that he had won the Nobel Prize for chemistry in 1944 and that nuclear bombs had been developed from his basic discovery. Later, as director of the Max-Planck-Gesellschaft (the postwar successor to the Kaiser Wilhelm Gesellschaft), he spoke vigorously against the misuse of atomic energy. Meitner—who many thought should have received the Nobel Prize with Hahn—continued to do nuclear research in Sweden and then England. Strassmann nurtured the study of nuclear chemistry in Mainz, Germany.

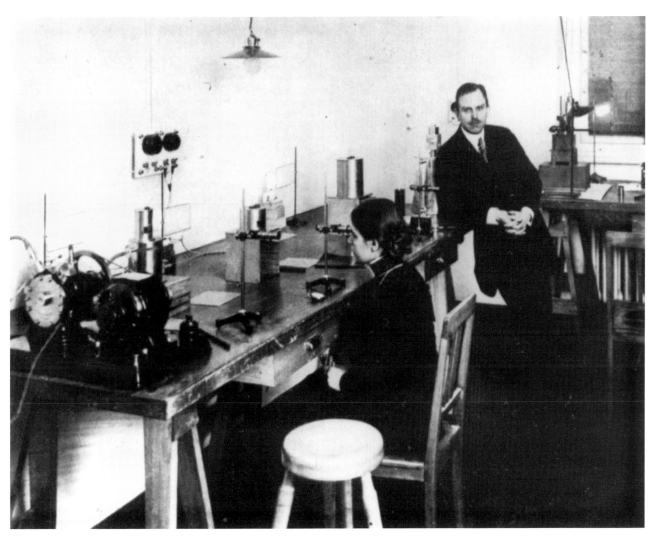

■ Lise Meitner and Otto Hahn in their laboratory.

■ Lise Meitner. Photograph Lotte Meitner-Graf. Courtesy Anne Meitner.

■ Otto Hahn. Courtesy Edgar Fahs Smith Memorial Collection, Department of Special Collections, University of Pennsylvania Library.

■ Fritz Strassmann. Courtesy Max–Planck–Gesellschaft, Munich.

GLENN THEODORE SEABORG (1912–1999)

Glenn Theodore Seaborg was involved in identifying nine transuranium elements (94 through 102), and he served as chairman of the United States Atomic Energy Commission (AEC) from 1961 to 1971. In 1951 he shared the Nobel Prize in chemistry with the physicist Edwin M. McMillan.

Born in Michigan, Seaborg earned his bachelor's degree from the University of California at Los Angeles and his doctorate in chemistry from the University of California at Berkeley. He then served as research assistant to Gilbert Newton Lewis (see Lewis, Chapter 6) and eventually became chancellor of the university. He worked away from Berkeley during two significant periods: once to participate in the Manhattan Project at the University of Chicago from 1942 to 1946 and then again to chair the AEC—from which he returned to Berkeley.

In 1940 Edwin McMillan, assisted by Philip Abelson (later editor of *Science* magazine), confirmed and elucidated the phenomenon of nuclear fission announced by Otto Hahn and Fritz Strassmann (see earlier in this chapter) in 1939. Specifically, he identified element 93, neptunium, among the fission products of uranium that was bombarded with neutrons produced from deuterons using the small (27-inch) cyclotron at Berkeley. McMillan also predicted the existence of element 94, plutonium, which he expected to find among the products of uranium under direct deuteron bombardment. McMillan, however, was suddenly called away to do war work and eventually joined the program at Los Alamos to build nuclear bombs. After World War II, his scientific reputation was enhanced by his critical contributions to the theory of particle accelerators.

Seaborg and his associates, who took over McMillan's project, soon found plutonium with a mass number of 238. Further research led to the production of isotope 239 in early 1941 in very small quantities. Plutonium 239 was shown to be fissionable by bombardment with slow neutrons and therefore became the newest material from which a nuclear bomb could be constructed.

■ Glenn T. Seaborg and President John F. Kennedy at Germantown, Maryland, headquarters of the Atomic Energy Commission, 16 February 1961. Courtesy Ernest Orlando Lawrence Berkeley Laboratory.

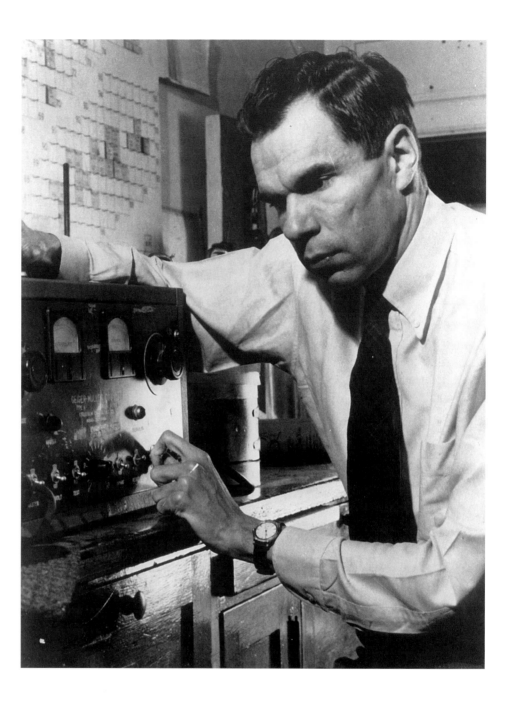

■ Glenn T. Seaborg in 1942, adjusting a Geiger counter. Courtesy Ernest Orlando Lawrence Berkeley Laboratory.

Up to that time scientists had known only of uranium 235 for this purpose. Seaborg then joined the Manhattan Project to work on the plan for producing sufficient plutonium 239 for a bomb—the one that was dropped on Nagasaki. Even before the war ended, he turned his attention to the production of further transuranium elements, developing the actinide transition series in the periodic table.

At the AEC, Seaborg became deeply involved in both arms control and nuclear regulatory affairs—attempting to manage the power of the atomic nucleus that his scientific work had revealed. Among chemists he has been unusual in writing histories of the epic developments in which he was involved so that the public can be the wiser for his experiences. With Benjamin S. Loeb he has written a historical series, the first of which was *Kennedy, Khrushchev, and the Test Ban* (1981).

6. Chemical Synthesis, Structure, and Bonding

Before scientists began to explore the internal structure of atoms, they were concerned about how and why compounds form. Berzelius's dualistic theory, for example, proposed that compounds are formed from atoms because of opposite electrical charges (see Berzelius, Chapter 3). But this theory did not explain the enormous number of compounds that were formed from four elements alone—carbon, oxygen, hydrogen, and nitrogen. Berzelius called them "organic" compounds because they always seemed to be the products of living beings composed of complex yet highly organized systems. The thinking was that such substances could not be created in the laboratory from inorganic materials, and thus a "vital force"—beyond the understanding of chemists— was necessary to explain their existence.

JUSTUS VON LIEBIG (1803–1873) AND FRIEDRICH WÖHLER (1800–1882)

Justus von Liebig and Friedrich Wöhler were friends who helped make organic chemistry a field of systematic study within the framework of known chemical laws. Years after their work, the insights gained from organic chemistry about how atoms bond were applied to the whole of chemistry.

Liebig learned to perform chemical operations as a child in his father's small laboratory, which was maintained to support the family drug and painting-materials business in Darmstadt, Germany. After Liebig finished his university studies in Germany, his ambitions led him to work in Paris with Gay-Lussac, who was in the forefront of chemical research at that time. Liebig was soon appointed to the University of Giessen, where he immediately set about providing the kind of opportunities he had enjoyed in Gay-Lussac's labora-

tory, but for many more students at a time. He thus created a model laboratory for training graduate students that was widely imitated in Europe and later on in the United States. From Giessen he also edited the journal that was to become the preeminent publication in chemistry—*Annalen der Chemie und Pharmacie*. After twenty-five years at Giessen he moved on to the University of Munich.

Wöhler, driven by a need similar to Liebig's to obtain the finest education in chemistry, went to Sweden to study with Berzelius after taking his medical degree at the University of Heidelberg. Even after returning to Germany, Wöhler remained Berzelius's loyal supporter for many years, translating several editions of Berzelius's textbook (see Berzelius, Chapter 3) into German as well as his annual reports on chemical

■ Justus von Liebig. Courtesy Edgar Fahs Smith Memorial Collection, Department of Special Collections, University of Pennsylvania Library.

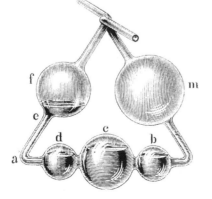

■ Liebig's five-bulbed apparatus, seen lying on the table in the portrait photograph. It was filled with a solution of caustic potash and used to collect carbon dioxide after burning a weighed sample to determine the percentage of carbon. Over the years it became the insignia of Liebig's students. From H. von Fehling, *Neues Hand-worterbuch der Chemie* (Braunschweig, 1875), p. 468.

■ Friedrich Wöhler. *Courtesy Edgar Fahs Smith Memorial Collection,
Department of Special Collections, University of Pennsylvania Library.*

progress. The University of Göttingen, where Wöhler taught for nearly fifty years, became—like Giessen with Liebig—an international mecca for chemistry graduate students.

The friendship between Liebig and Wöhler began in 1825 after they amicably resolved a dispute over two substances that had apparently the same composition—cyanic acid and fulminic acid—but very different characteristics: The silver compound of fulminic acid, investigated by Liebig, was explosive, whereas silver cyanate, as Wöhler found, was not. These and similar substances, called "isomers" by Berzelius, led chemists to suspect that substances are defined not simply by the number and kind of atoms in the molecule but also by the *arrangement* of those atoms.

Perhaps the most famous creation of an isomeric compound was Wöhler's accidental synthesis of urea in 1828, when he was attempting to prepare ammonium cyanate (which he later succeeded in preparing by allowing the crystals to form at room temperature instead of by evaporating the solution). The feat of imitating nature in the laboratory was a truly exciting experience—as Wöhler expressed it in his often-quoted letter to Berzelius: "I can no longer, so to speak, hold my chemical water and must tell you that I can make urea without needing a kidney, whether of man or dog; the ammonium salt of cyanic acid is urea."

The most significant result of Wöhler's and Liebig's collaboration was their discovery of certain stable groupings of atoms in organic compounds that retain their identity, even when those compounds are transformed into others. The first such grouping they identified was the "benzoyl radical," found in 1832 during a study of oil of bitter almonds (benzaldehyde) and its derivatives. Their original objective was to buttress Berzelius's dualism theory in the realm of organic chemistry by thinking of radicals as organic chemical equivalents of inorganic atoms. But they gradually recognized that the substitutions that chemists effected within radicals—of, for example, electropositive hydrogen by electronegative chlorine—seriously threatened dualism as a comprehensive explanation of bonding in organic chemistry. In the long run their identification of radicals can be seen as an early step along the path to structural chemistry.

In the 1840s Liebig and Wöhler moved away from fundamental research in organic chemistry. Among Liebig's new passions were agricultural chemistry and physiology—interests that influenced a number of his American students, who founded agricultural experimental stations and agricultural education in the United States. Wöhler returned to his early inorganic chemistry interests, having successfully extracted aluminum and beryllium from their compounds by chemical means in 1827, the same year he synthesized urea. Among other contributions he prepared calcium carbide and discovered various silicon compounds, demonstrating close analogies to the chemistry of carbon.

Be-Liebig-es

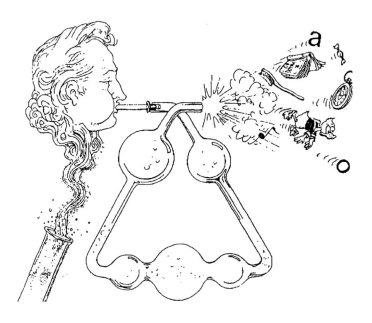

■ Cartoon of the Liebig apparatus. "Beliebiges" means "whatever"—presumably a chemist could pass any gaseous matter through the potash-filled bulbs to remove carbon dioxide. Cartoon by Rolf Hank de Vries for *Liebigschule Giessen Festschrift zur 150-Jahr-Feier* (1987), edited by Erwin Glaum.

■ Justus von Liebig's laboratory in Giessen around 1840, drawn by Wilhelm von Trautschold. Among Liebig's students depicted here is August W. Hofmann, William Perkin's professor (far right). Courtesy Liebig Museum, Giessen, Germany.

WILLIAM HENRY PERKIN (1838–1907)

In 1856, during Easter vacation from the Royal College of Chemistry, eighteen-year-old William Henry Perkin synthesized mauve, or aniline purple—the first synthetic dyestuff—from chemicals derived from coal tar. Like Wöhler's accidental synthesis of urea, Perkin's chemical manipulations were designed to produce a quite different product—quinine. His teacher, August W. Hofmann, one of Liebig's former students, had remarked on the desirability of synthesizing this antimalarial drug, which at that time was derived solely from the bark of the cinchona tree, by then grown mainly on plantations in southeast Asia. Against Hofmann's recommendation and with the financial support of his father, a construction contractor, Perkin commercialized his discovery and developed the processes for the production and use of the new dye. In 1857 he opened his factory at Greenford Green, not far from London.

From this modest beginning grew the highly innovative chemical industry of synthetic dyestuffs and its near relative, the pharmaceutical industry, which improved the quality of life for the general population. These two industries also stimulated the search for a better understanding of the structure of molecules. Perkin, at the age of thirty-six, sold his business so that he could devote himself entirely to research, which included early investigations of the ability of some organic chemicals to rotate plane-polarized light, a property used in considering questions of molecular structure.

■ A photograph that William Henry Perkin took of himself at the age of fourteen—four years before he discovered the first synthetic dyestuff.

■ William Henry Perkin holding a skein dyed mauve for a portrait painted by Arthur Cope in honor of the fiftieth anniversary of aniline dyes. In *Jubilee of the Discovery of Mauve and of the Coal-Tar Colour Industry by Sir W. H. Perkin*. Edited by Raphael Meldola, Arthur G. Green, and John Cannel Cain. London: Perkin Memorial Committee, 1906.

■ Archibald Scott Couper in Paris in 1857 or 1858. Courtesy Edgar Fahs Smith Memorial Collection, Department of Special Collections, University of Pennsylvania Library.

ARCHIBALD SCOTT COUPER (1831–1892) AND
AUGUST KEKULÉ VON STRADONITZ (1829–1896)

In 1858 Archibald Scott Couper and August Kekulé von Stradonitz, two young men from different backgrounds—and, as it turned out, entering upon even more disparate career paths—independently recognized that carbon atoms can link directly to one another to form carbon chains. This finding explained the very multiplicity of carbon compounds that had been puzzling chemists. The discovery by these two scientists depended on Kekulé's theory, proposed in 1857, that carbon is tetravalent—valence being defined at the time as the combining capacity of the elements. Couper, in his paper—and in another paper on salicylic acid that appeared earlier in 1858—indicated valence bonds as straight lines linking the symbols for the elements, which is still the practice in most modern structural diagrams.

Couper, who was a Scot educated in Glasgow, Edinburgh, Berlin, and Paris, came to chemistry from the study of philosophy and classical languages. This background probably helped him make an analogy between letters in words and carbon atoms in molecules, and focus on how carbon atoms combine with other atoms. His invention of an appropriate symbolic language to indicate the order in which the various atoms are joined in molecules may also stem from his philosophical and linguistic training. Sadly, his paper describing carbon linkage was read before the French Academy a few weeks after Kekulé's similar paper was published in Liebig's *Annalen*. Couper had entrusted his paper to Charles Adolphe Wurtz, in whose laboratory he worked in Paris, and Wurtz had procrastinated in giving it to an

■ Archibald Scott Couper's bond lines in a French version of his 1858 paper. On the left is his representation of tartaric acid and the product obtained after loss of water by heating. On the right is the first depiction of a ring system—for cyanuric acid (Az = N). Here Couper used continuous lines and brackets to represent bonds. In other publications, bonds are straight dotted lines—possibly the typesetter's preference. From *Annales de chimie et de physique*. Série 3, 53 (1858), 488–489.

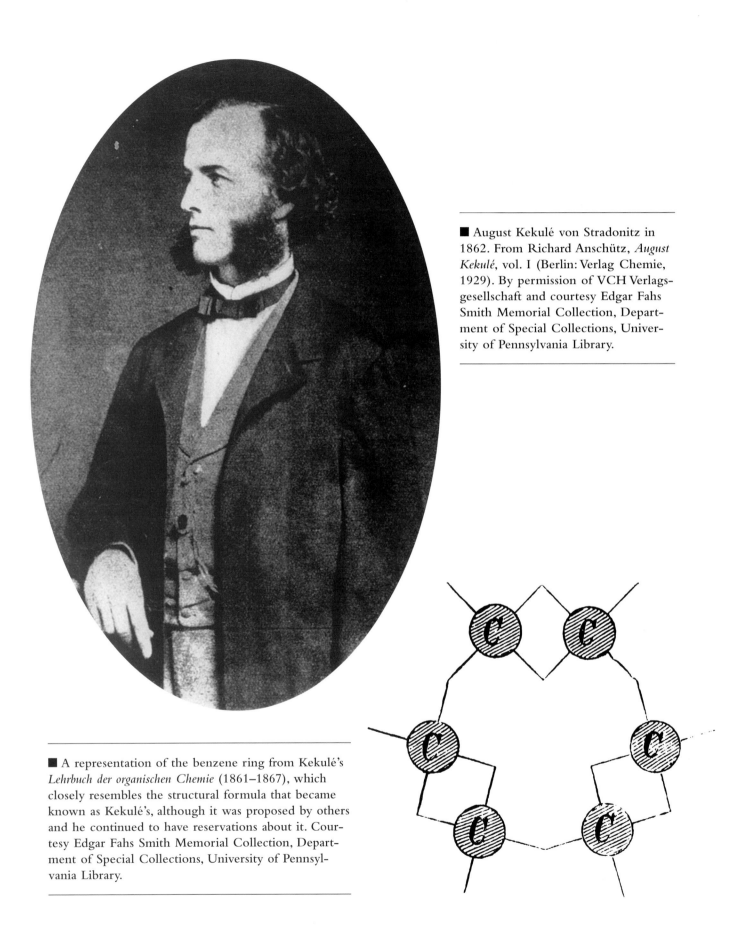

■ August Kekulé von Stradonitz in 1862. From Richard Anschütz, *August Kekulé*, vol. I (Berlin: Verlag Chemie, 1929). By permission of VCH Verlagsgesellschaft and courtesy Edgar Fahs Smith Memorial Collection, Department of Special Collections, University of Pennsylvania Library.

■ A representation of the benzene ring from Kekulé's *Lehrbuch der organischen Chemie* (1861–1867), which closely resembles the structural formula that became known as Kekulé's, although it was proposed by others and he continued to have reservations about it. Courtesy Edgar Fahs Smith Memorial Collection, Department of Special Collections, University of Pennsylvania Library.

Academy member for presentation. It is not known how much Couper's bitterness over his loss of priority and his subsequent fight with Wurtz contributed to his emotional collapse. He soon retreated to his Scottish home and never published another scientific paper for the remaining thirty years of his life.

Kekulé, a German of Czech descent, was intended by his family to become an architect, but at the University of Giessen he was lured to chemistry by Liebig's lectures. During his studies he also worked in Paris with Charles Gerhardt, another leading chemist in the effort to understand the constitution of organic compounds. Kekulé, after receiving his doctorate from Giessen, served as a research assistant, first in Switzerland, then in England. On a bus ride in London on the way home from visiting a chemist friend, Kekulé envisioned his earliest notion of carbon chains. In a daydream he "saw" carbon atoms joining in a "giddy dance."

Kekulé returned to Germany from England to begin his career as a university teacher, which took him from the University of Heidelberg, to the University of Ghent in Belgium, and finally to the University of Bonn, where he oversaw the establishment of a chemical institute. He was a good lecturer, well liked by his many students. As with several other scientists in chemical history, writing a textbook—*Lehrbuch der organischen Chemie*—proved to be a stimulus for new chemical theories. One afternoon in 1865 in Ghent, while he was working on his textbook, he became sleepy and turned his chair to the fire. Again he saw dancing strings of carbon atoms, but this time he saw one that snake-like took its tail into its mouth, which gave him the idea for the ring form of the benzene molecule. Here then was a structure for the many molecules that would not fit into the existent structural theory.

■ Heinrich von Angeli's portrait of August Kekulé von Stradonitz, commissioned by German dye companies for his sixtieth birthday. Courtesy Edgar Fahs Smith Memorial Collection, Department of Special Collections, University of Pennsylvania Library.

JACOBUS HENRICUS VAN'T HOFF (1847–1930)

In 1872 Jacobus Henricus van't Hoff, a Dutch graduate student, came to Bonn to study for a year. From Kekulé, van't Hoff learned of a possible tetrahedral arrangement of the valence bonds of carbon, proposed by the Russian chemist Alexander Mikhailovich Butlerov in 1862. In 1873, after van't Hoff had moved to Paris to work with Charles Adolphe Wurtz, he realized that the phenomenon of optical activity possessed by some organic molecules—their ability to rotate plane-polarized light—could be explained by the two possible arrangements of four different substituents in the space around a carbon atom. This theory provided substantial indication that the molecular structures the chemists of the time were discussing had a physical reality in three-dimensional space and were not just aids to conceptualizing molecules. (Another graduate student working

■ A young Jacobus Henricus van't Hoff. Courtesy Edgar Fahs Smith Memorial Collection, Department of Special Collections, University of Pennsylvania Library.

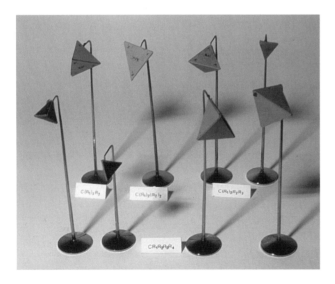

■ Van't Hoff disseminated his stereochemical ideas to leading chemists of the day by sending them three-dimensional paper models of tetrahedral molecules, like these now housed in the Leiden Museum. Photograph courtesy O. Bertrand Ramsay.

in Wurtz's laboratory, Joseph Achille Le Bel, arrived at the same explanation of optical activity independently.)

Van't Hoff returned to the Netherlands to complete his doctoral degree. He was soon appointed lecturer in theoretical and physical chemistry at the University of Amsterdam, where he stayed for twenty years. There he conducted the studies of reaction rates, chemical equilibrium, chemical affinity, and osmotic pressures that helped found the discipline of physical chemistry. In 1896 he moved to the University of Berlin, and in 1901 he became the first Nobel laureate in chemistry for his work in physical chemistry.

■ Jacobus Henricus Van't Hoff in 1904. Courtesy Edgar Fahs Smith Memorial Collection, Department of Special Collections, University of Pennsylvania Library.

GILBERT NEWTON LEWIS (1875–1946) AND IRVING LANGMUIR (1881–1957)

Once physicists studying the structure of the atom began to realize that the electrons surrounding the nucleus had a special arrangement, chemists began to investigate how these theories corresponded to the known chemistry of the elements and their bonding abilities. Two Americans who were instrumental in developing a bonding theory based on the number of electrons in the outermost "valence" shell of the atom were Gilbert Newton Lewis and Irving Langmuir.

In 1902, while Lewis was trying to explain valence to his students, he depicted atoms as constructed of a concentric series of cubes with electrons at each corner. This "cubic atom" explained the eight groups in the periodic table and represented his theory that chemical bonds are formed by electron transference to give each atom a complete set of eight. In 1923 he redefined acids as any atom or molecule with an incomplete "octet" that were thus capable of accepting electrons from another atom; bases were, of course, electron donors.

Lewis was also important in developing the field of thermodynamics and applying its laws to real chemical systems. At the end of the nineteenth century when he started working, the law of conservation of energy and other thermodynamic relations were known only as isolated equations. Lewis built on the work of another American pioneer in thermodynamics, Josiah Willard Gibbs (1839–1903) of Yale University, whose contributions were only slowly recognized. Their work was of immense value in predicting whether reactions will go almost to completion, reach an equilibrium, or proceed almost not at all, and whether a mixture of chemicals can be separated by distillation.

Lewis was educated at home, while his family lived in Massachusetts and Nebraska, until he was fourteen years old. His subsequent education was more conventional, although nonetheless stimulating, and included a Ph.D. from Harvard University earned under Theodore W. Richards (see Richards, Chapter 4). Lewis then made the pilgrimage to Germany to work with the physical chemists Walther Nernst and Wilhelm Ostwald. He held several university faculty appointments, including ones at the Massachusetts Institute of Technology and the University of California at Berkeley, where he expanded the programs in chemistry and chemical engineering.

As a research pioneer for the General Electric Company, Irving Langmuir made scientific contributions in chemistry, physics, and atmospheric science. He received his doctorate from Walther Nernst in Göttingen, Germany, but became bored after one year of teaching college. In 1909 he arrived at the recently established GE Research Laboratory. His first job was to solve the

---

■ Gilbert Newton Lewis's memorandum of 1902 showing his speculations about the role of electrons in atomic structure. From *Valence and the Structure of Atoms and Molecules* (1923), p. 29.

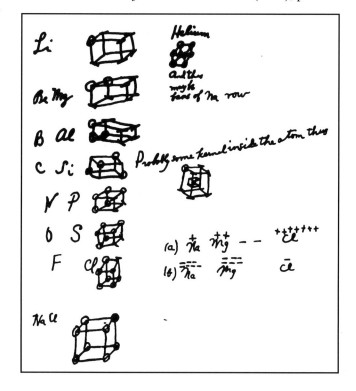

problems they were having with the new tungsten filament light bulbs. Langmuir concentrated on the basic principles on which the lamp operated, investigating the chemical reactions catalyzed by the hot tungsten filament. He suggested filling the bulbs with nitrogen gas (and later argon gas) and twisting the filament into a spiral form to inhibit the vaporization of tungsten.

His interest in fundamentals involved him in the theory of chemical bonding in terms of electrons, and he elaborated on ideas first expressed by Gilbert Lewis. Langmuir proposed that octets could be filled by sharing pairs between two atoms—the "covalent" bond. His studies of surface chemistry—the study of chemical forces at the contact surfaces (interfaces) between different substances, where so many biologically and technologically important reactions occur—earned him the Nobel Prize in chemistry in 1932. Langmuir developed a new concept of adsorption, according to which every molecule striking a surface remains in contact with it before evaporating, thus forming a firmly held monolayer—in contrast to earlier theories that likened adsorption to the attraction of the earth for the gases in the atmosphere, where the attraction diminishes as distance from the earth increases. He developed a multitude of experimental techniques, including the extensive use of vacuum tubes to study solid–gas interfaces and of oil films to study liquid–liquid interfaces. Other practical work with theoretical implications—on electrical discharges in gases—helped to lay the foundation for "plasma" physics, which has application today in attempts at controlled nuclear fusion. He maintained a lifelong interest in meteorology, including work developing aircraft de-icing capabilities during World War II.

Here too Langmuir pushed beyond observation to theory, which led to his carrying out early experiments in "seeding" clouds with solid carbon dioxide particles to produce rain.

■ Gilbert Newton Lewis using a slide rule at his desk.

■ Gilbert Newton Lewis. Courtesy The Chemists' Club.

■ Irving Langmuir at home enjoying *Harper's Magazine*.
Courtesy The Chemists' Club.

■ One of Linus Pauling's many peacekeeping activities—flipping pancakes at the 1963 pancake breakfast of the Pasadena branch of the Women's International League for Peace and Freedom. At right, his wife, Ava Helen Pauling. Courtesy WILPF.

LINUS CARL PAULING (1901–1994)

The quantum mechanics developed in the 1920s was quickly applied to problems of bonding and structure, which were the focus of much of the prolific research of Linus Carl Pauling and earned him the Nobel Prize in chemistry in 1954.

Born in Portland, Oregon, Pauling received his bachelor's degree from Oregon State University and his Ph.D. in chemistry from California Institute of Technology—despite the economic difficulties experienced by his family. After a European sojourn investigating the implications of the new quantum mechanics for chemistry, mainly in Munich with the theoretical physicist Arnold Sommerfeld, he joined the faculty at Caltech, where he remained until 1963. Then, after a short stay at the Center for the Study of Democratic Institutions in Santa Barbara, California, he resumed his laboratory research at the University of California at San Diego, from which he moved on to Stanford University, and finally to the Linus Pauling Institute of Science and Medicine in Palo Alto.

Among his accomplishments, he determined crystal structures by X-ray crystallography and the structures of gas molecules by electron diffraction. He studied the magnetic properties of substances, including hemoglobin, which helped him ascertain the molecular cause of sickle-cell anemia. He developed an electronegativity scale to assign to atoms involved in covalent and ionic bonding, and he formulated the concept of "resonance" to talk about the state of a chemical system where none of the classical structural formulas is entirely consistent with observed properties. He extended the theory of covalent bonds to include metals and intermetallic compounds. He proposed helical structures for proteins based on the coplanarity of the atoms in the peptide bond. But Pauling is perhaps best known to the public for championing the use of vitamin C to maintain and restore health.

In the 1950s and afterward he campaigned tirelessly—and in the face of significant professional and governmental opposition—to put an end to nuclear bomb tests in the atmosphere and to the arms race. In 1963, the year that the Nuclear Test Ban Treaty went into effect, Pauling was given the Nobel Peace Prize, the second person after Marie Curie to win a second Nobel Prize.

■ Linus Pauling with molecular models. Courtesy Archives, California Institute of Technology.

7. Pharmaceuticals
and the
Path to Biomolecules

In 1906 Paul Ehrlich prophesied the role of modern-day pharmaceutical research, predicting that chemists in their laboratories would soon be able to produce substances that would seek out specific disease-causing microorganisms—"magic bullets," as he called them. Ehrlich himself met with signal successes in the emerging fields of serum antitoxins and in chemotherapy. The achievers in this chapter—both those who elucidated the structures of complex organic molecules and those who used that knowledge to synthesize them—helped create many of the medical "miracles" that improve our quality of life today.

PAUL EHRLICH (1854–1915)

Paul Ehrlich was born near Breslau—then in Germany, but now known as Wrocław, Poland—and studied to become a medical doctor at the university there and in Strasbourg, Freiburg im Breisgau, and Leipzig. In Breslau he worked in the laboratory of his cousin, Carl Weigert, a pathologist who pioneered the use of aniline dyes as biological stains. Ehrlich became interested in the selectivity of dyes for specific organs, tissues, and cells, and he continued his investigations at the Charité Hospital in Berlin. After he showed that dyes react specifically with various components of blood cells and the cells of other tissues, he began to test the dyes for therapeutic properties to determine whether they could kill off pathogenic microbes.

After his own bout with tuberculosis—probably contracted in the laboratory—and his subsequent cure with Robert Koch's tuberculin therapy, Ehrlich focused his attention on bacterial toxins and antitoxins. At first he worked in a small private laboratory, but as the quality of his work became recognized by Koch and others,

he was able to command more and better resources—eventually the State Serum Institute in Frankfurt. In 1908 he received the Nobel Prize in medicine for his work on immunization.

In Frankfurt he continued to look for chemical agents to use against disease. He obtained the cooperation of the nearby Cassella chemical works, which donated samples of new compounds produced in their laboratories for him to test for biological activity. In 1906 Georg-Speyer-Haus, a research institute for chemotherapy, was established with its own staff under Ehrlich's direction. The research programs were guided in part by Ehrlich's theory that the germicidal capability of a molecule depended on its structure, particularly its side-chains, which could bind to the disease-causing organism. The most successful products of this quest were Salvarsan (1912)—dihydroxydiaminoarsenobenzenedihydrochloride—and Neosalvarsan (1912), the most effective drugs for treating syphilis until the advent of antibiotics in the 1940s.

■ Paul Ehrlich in his study about 1910.
Courtesy ECON Verlag GmbH.

■ William Lawrence Bragg (left) and William Henry Bragg. Courtesy Edgar Fahs Smith Memorial Collection, Department of Special Collections, University of Pennsylvania Library.

WILLIAM HENRY BRAGG (1862–1942) AND WILLIAM LAWRENCE BRAGG (1890–1971)

For progress in pharmaceuticals research to occur along the lines Paul Ehrlich had suggested, the structure of complex organic molecules had to be studied in greater depth. By studying the chemical reactions that a compound and its degradation products could enter into with other compounds of known structure, chemists were able to deduce the structures of many complex organic molecules—although it sometimes took years of experimentation and analysis of results. When X-ray crystallography was introduced in 1912, it became possible to determine molecular structure from the compound itself. In this method, structural information is obtained by mathematical analysis of the intensity of X rays scattered (or diffracted) from parallel planes in a crystal, as recorded photographically or by an electronic detector. Until electronic computers were developed during World War II, these calculations were incredibly laborious. In 1915 a unique father-son team, William Henry Bragg and his son, William Lawrence Bragg, won the Nobel Prize in physics for their seminal roles in X-ray crystallography.

William Henry's mother died when he was just seven years old, and he was sent to live with a bachelor uncle, whose place he regarded as home during his years away at school and at Cambridge University. At the university he studied mathematics, but his first academic appointment, to the University of Adelaide in Australia—in mathematics and physics—required that he learn much of physics on his own.

Not until 1903–1904, at the age of forty-one, did William Henry embark on studies of ionizing radiation—a rather late-in-life start on a topic that was to bring a Nobel Prize. On the basis of his early publications on radiation, he was named Cavendish professor at the University of Leeds in 1908. His eldest son, William Lawrence, who had begun his university studies in mathematics in Australia, transferred to Cambridge, where he changed his focus to physics.

In 1912 Max von Laue reported the diffraction of X rays by a crystal (for which he received a Nobel Prize in physics in 1914). The elder Bragg and his son, who was by then a doctoral student with J. J. Thomson (see Thomson, Chapter 5) at Cambridge, began exploring this phenomenon immediately. They brought different interests and skills to the collaboration. William Henry's original interest was in what diffraction showed about the nature of X rays, and he was a skilled experimenter and designer of instruments. William Lawrence was more concerned with what X rays revealed about the crystalline state, and he possessed a powerful ability to conceptualize physical problems and express them mathematically. Simple inorganic crystals like sodium chloride were the subjects of the original studies in X-ray crystallography. Here the surprising result was that in the solid state these ionic compounds did not exist as paired positive and negative ions. Sodium chloride, for instance, did not exist as NaCl units; rather, Na and Cl alternated in a regular fashion in the crystal lattice.

In 1915, the same year William Henry received the Nobel Prize, he was appointed to University College, London. But the work on X-ray crystallography came to a halt during World War I, and both Braggs served as scientific advisers to the military—especially on the problem of submarine detection. After World War I, William Lawrence began his academic career, following in the footsteps of his fellow Australian Ernest Rutherford—first at the University of Manchester and then at the Cavendish Laboratory in Cambridge. In 1923 William Henry became head of the Royal Institution, a position in which he served twenty years and to which his son succeeded in 1954. In their various positions the Braggs continued their work in X-ray crystallography and built up programs for doctoral and postdoctoral students. Under their leadership the field moved on to such fields of study as the structure of metals and organic compounds and later to those of biochemical and pharmaceutical importance.

■ An early Bragg X-ray spectrometer. *L*, lead box; *A*, *B*, *D*, slits; *C*, crystal; *I*, ionization chamber; *V*, vernier of crystal table; *V'*, vernier of ionization chamber; *K*, earthing key; *E*, electroscope; *M*, microscope.

DOROTHY CROWFOOT HODGKIN (1910–1994)

Among the X-ray crystallographers inspired by the Braggs was Dorothy Crowfoot Hodgkin, the third woman ever to win the Nobel Prize in chemistry, which she received in 1964.

Dorothy Crowfoot was born in Cairo, Egypt, to English parents. Although her formal schooling took place in England, she spent a significant part of her youth in the Middle East and North Africa, where her father was a school inspector. Both her parents were authorities in archaeology, and she almost followed the family vocation; but from childhood she was fascinated by minerals and crystals. She enjoyed using a portable mineral analysis kit given to her when she became interested in analyzing pebbles she and her sister found in the stream that ran through the Crowfoot's garden in Khartoum, Sudan. When she was fifteen, her mother gave her Sir William Henry Bragg's *Concerning the Nature of Things* (1925), which contained intriguing discussions of how scientists could use X rays to "see" atoms and molecules.

At Somerville College, Oxford, Crowfoot studied physics and chemistry and chose to do her fourth-year research project on X-ray crystallography, a field for which the equipment at Oxford was primitive. After graduation she seized the opportunity of studying at Cambridge with John Desmond Bernal, who had worked for five years with W. H. Bragg at the Royal Institution. Bernal and Crowfoot collaborated successfully, using X-ray crystallography to elucidate the three-dimensional structure of complex and biologically important molecules, including the sterols—complex alcohols found in plant and animal tissues that are related to hormones—and pepsin—the digestive enzyme

that was the first protein to be so analyzed. In 1937 Crowfoot received her Ph.D. from Cambridge—the same year that she married Thomas L. Hodgkin, who became an authority on African history. Both Hodgkins held academic appointments at Oxford, and they raised their three children there with the help of the Hodgkin grandparents.

Dorothy Hodgkin's three greatest chemical achievements were her determination of the structures of penicillin—part of the Anglo-American program to synthesize this new antibiotic during World War II (1945); of vitamin B_{12}, the essential vitamin that prevents pernicious anemia (1957); and of insulin, the hormone essential for successful carbohydrate metabolism—a puzzle on which she worked sporadically from 1934 to 1972.

Hodgkin is fondly remembered by her group of research students, which included many women. She was also involved in a wide range of peace and humanitarian causes and was especially concerned for the welfare of scientists and people living in nations defined by the United States and the United Kingdom as adversaries in the 1960s and 1970s—for example, the Soviet Union, China, and North Vietnam. From 1976 to 1988 she was chair of the Pugwash movement, which was originally inspired by the concerns voiced in 1955 by Albert Einstein and the philosopher-mathematician Bertrand Russell that work by scientists—such as the creation of the hydrogen bomb—would lead to conflict and needed the insights of and input from the world's scientists. Later the Pugwash conferences dealt with other potential dangers raised by scientific research.

■ Dorothy Crowfoot Hodgkin. Copyright 1967, The Nobel Prize Foundation.

PERCY LAVON JULIAN (1899–1975) AND CARL DJERASSI (1923–)

In the 1930s chemists recognized the structural similarity of a large group of natural substances, including cholesterol, bile acids, sex hormones, and the cortical hormones of the adrenal glands—the steroids. The medicinal potential of various steroids quickly became obvious, but extracting sufficient quantities of these substances, which exist in minute amounts in animal tissue and fluids, was prohibitively expensive. As with other scarce or difficult-to-isolate natural products like quinine or penicillin, chemists were called upon to mimic nature—to create the substances synthetically—and later to modify the natural molecule to make the substance safer and more effective as a drug. For steroids, chemists found starting materials in certain plant substances that were also steroids. Percy Lavon Julian and Carl Djerassi participated actively in the synthesis of natural steroids and their large-scale production.

Julian was born in Montgomery, Alabama, in 1899—the son of a railway mail clerk and the grandson of slaves. In an era when African Americans faced prejudice in virtually all aspects of life—not least in the scientific world—he succeeded against the odds. He majored in chemistry at De Pauw University, earning his way by digging ditches and by waiting on tables in a fraternity house. After graduation he worked at Fisk University for two years as a chemistry instructor. He then completed a master's degree in organic chemistry at Harvard University and returned to teaching at West Virginia State College. In 1929 Julian began his studies at the University of Vienna, focusing on the chemistry of medicinal plants, and earned his doctorate in 1931. He and a fellow doctorate, Josef Pikl (whom Julian had invited and assisted in coming to America), took positions at Howard University and two years later moved to Julian's alma mater, De Pauw. There they accomplished the first total synthesis of the active principle of the Calabar bean—an alkaloid, physostigmine, used in the treatment of glaucoma.

Meanwhile German chemists were showing that the

■ Percy Julian as a DePauw University student. Courtesy DePauw University Archives and Special Collections.

steroid stigmasterol—which Julian had obtained as a by-product of the physostigmine synthesis but was also obtainable from soybeans—could be used in the preparation of certain sex hormones. In pursuit of this lead, in 1936 Julian wrote to the Glidden Company in Chicago, requesting some sample soybean oil. The Glidden vice president, W. J. O'Brien, was also on the board of the Institute of Paper Chemistry in Appleton, Wisconsin, where Julian had recently applied for a job as research chemist. After listening to his fellow board members bemoan the difficulty in hiring Julian because

■ Percy Lavon Julian. Gift of Ray Dawson.

of a law forbidding Negroes to stay overnight in Appleton, O'Brien decided to offer him a job at Glidden. "If he is half as good as they say he is," said O'Brien, "I can use him at Glidden." Julian was promptly made Director of Research of the Glidden Soya Products Division, where he remained until 1957, when he founded his own company, Julian Laboratories.

Julian and Djerassi were both involved in the exciting competition of the late 1940s and early 1950s to synthesize the hormone cortisone inexpensively. In 1949 Julian published a paper on a new synthesis for Reichstein's Substance S, which is also present in the

adrenal cortex and differs from cortisone in lacking only one oxygen atom in a particular molecular position. Hydrocortisone is still widely produced from this substance.

Djerassi came to the United States in 1939, after fleeing from the Nazis in Austria. He sped through college, after which he spent a brief period working on antihistamines for the Swiss pharmaceutical company CIBA at their New Jersey facility. He then completed a doctorate in organic chemistry at the University of Wisconsin, where he wrote his dissertation on how to transform the male sex hormone testosterone into the female

■ The announcement of a new cortisone synthesis by Syntex in 1951. Carl Djerassi is seated third from the right at the table, in the dark suit. On the table sits a huge Mexican yam, the source of the starting material for Syntex's cortisone and for norethindrone. Courtesy Carl Djerassi.

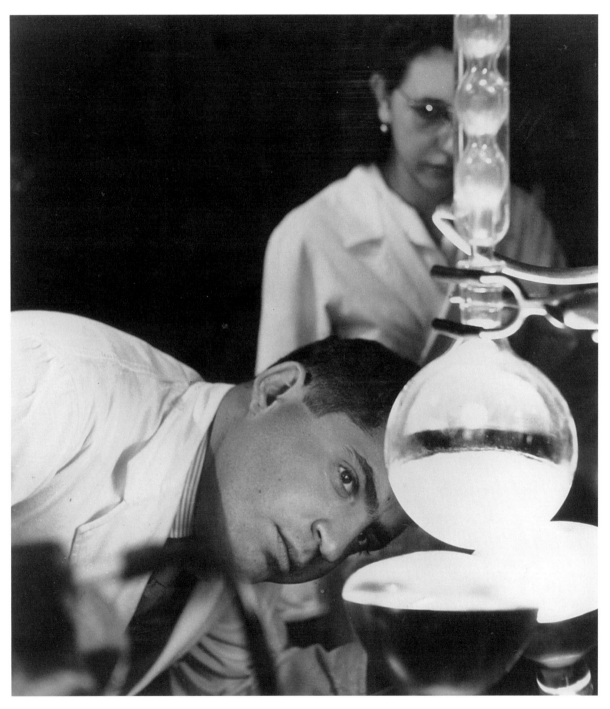

■ Carl Djerassi at Syntex in 1951, with his assistant Arelina Gonzalez, working on the synthesis of norethindrone, the basis of the first effective oral contraceptive. Note that in those days chemists were less concerned about protecting their eyes. Courtesy Carl Djerassi.

sex hormone estradiol using a series of chemical reactions. Djerassi's longtime fascination with steroids prompted his return to CIBA, but he was not allowed to work on steroid synthesis there. That promising field of research was reserved for the laboratories at CIBA's corporate headquarters in Switzerland. Djerassi was disappointed, and in 1949—shortly after Julian's paper on Substance S appeared—he joined Laboratorios Syntex S.A. in Mexico City. The steroid chemist Russell Marker and a partner had founded the firm to produce synthetic hormones from steroidal substances in Mexican plants—the same reason that Julian founded Julian Laboratories. Syntex was trying to synthesize cortisone from diosgenin, a steroidal substance found in Mexican yams; the company had already produced male and female sex hormones from diosgenin. In 1951 Djerassi's group successfully synthesized cortisone, improving on the original procedure done in 1944 by Lewis Sarett of Merck: The Syntex process not only used a cheaper raw material but also required only about half as many steps.

In 1951, the same year that Djerassi's group synthesized cortisone, it also synthesized the first effective oral contraceptive. It had long been known that during pregnancy progesterone serves as a natural contraceptive by inhibiting further ovulation, while maintaining the proper uterine conditions for the fetus. But taking natural progesterone orally weakens its biological activity, and thus the search was on for a more active sex hormone that could survive digestive processes. A compound synthesized at Syntex proved to be one of the most potent oral progestins ever made. In 1957 the Food and Drug Administration approved Syntex's norethindrone, plus a related drug produced by G. D. Searle and Company, first as a treatment for menstrual difficulties and then as a birth-control pill.

Djerassi maintained a twenty-year-long relationship with Syntex, while also accepting academic appointments after his 1951 triumphs, first at Wayne State University in Detroit and then at Stanford University. He made many more advances in synthetic organic chemistry and refined the techniques of mass spectroscopy and methods for deducing the precise orientation in space of the atoms in a molecule from optical rotatory dispersion (ORD).

Both Julian and Djerassi placed a high priority on their contributions to society as scientists and citizens. Julian was particularly active in groups seeking to advance conditions for African Americans, helping to found the Legal Defense and Educational Fund of Chicago as well as serving on the boards of several other organizations and universities. He was always attempting to build bridges between diverse groups of people. Djerassi, in line with his work on the birth-control pill, seeks to raise consciousness about the global need for population control. Stemming from his years with Syntex have been his efforts to encourage science in developing countries like Mexico. In memory of his daughter, who was an artist, and consonant with his own artistic and literary interests, he has established a colony for artists near Santa Cruz, California.

JAMES WATSON (1928–), FRANCIS CRICK (1916–), MAURICE WILKINS (1916–), AND ROSALIND FRANKLIN (1920–1958)

In 1962 James Watson, Francis Crick, and Maurice Wilkins jointly received the Nobel Prize in medicine or physiology for their determination in 1953 of the structure of deoxyribonucleic acid (DNA). Because the Nobel Prize can be awarded only to the living, Wilkins's colleague Rosalind Franklin, who died from cancer at the age of thirty-seven, could not be honored.

The molecule that is the basis for heredity, DNA, contains the patterns for constructing proteins in the body, including the various enzymes. A new understanding of heredity and hereditary disease was possible once it was determined that DNA consists of two chains twisted around each other, or double helixes, of alternating phosphate and sugar groups and that the two chains are held together by hydrogen bonds between pairs of organic bases—adenine (A) with thymine (T) and guanine (G) with cytosine (C). Modern biotechnology also has its basis in the structural knowledge of DNA—in this case the scientist's ability to modify the DNA of host cells that will then produce a desired product, for example, insulin.

The background for the work of the four scientists was formed by several scientific breakthroughs: the progress made by X-ray crystallographers in studying organic macromolecules; the growing evidence supplied by geneticists that it was DNA, not protein, in chromosomes that was responsible for heredity; Erwin Chargaff's experimental finding that there are equal numbers of A and T bases and of G and C bases in DNA; and Linus Pauling's discovery that the molecules of some proteins have helical shapes—arrived at through the use of atomic models and a keen knowledge of the possible disposition of various atoms.

Of the four DNA researchers only Rosalind Franklin had any degrees in chemistry. The daughter of a prominent London banking family, where all children—girls and boys—were encouraged to develop their individual aptitudes, she held her undergraduate and graduate degrees from Cambridge University. During World War II, she gave up her research scholarship to contrib-

ute to the war effort at the British Coal Utilization Research Association, where she performed fundamental investigations on the properties of coal and graphite. After the war she joined the Laboratoire Centrale des Services Chimiques de l'Etat in Paris, where she was introduced to the technique of X-ray crystallography and rapidly became a respected authority in this field. In 1951 she returned to England to King's College, London, where her charge was to upgrade the X-ray crystallographic laboratory there for work with DNA.

Already at work at King's College was Maurice Wilkins, a New Zealand–born but Cambridge-educated physicist. As a new Ph.D. he worked during World War II on the improvement of cathode-ray tube screens for use in radar and then was shipped out to the United States to work on the Manhattan Project. Like many other nuclear physicists, he became disillusioned with his subject when it was applied to the creation of the atomic bomb; he turned instead to biophysics, working with his Cambridge mentor, John T. Randall—who had undergone a similar conversion—first at the University of St. Andrews in Scotland and then at King's College, London. It was Wilkins's idea to study DNA by X-ray crystallographic techniques, which he had already begun to implement when Franklin was appointed by Randall. The relationship between Wilkins and Franklin was unfortunately a poor one and probably slowed their progress.

Meanwhile, in 1951 twenty-three-year-old James Watson, a Chicago-born American, arrived at the Cavendish Laboratory in Cambridge. Watson had two degrees in zoology: a bachelor's from the University of Chicago and a doctorate from the University of Indiana, where he became interested in genetics. He worked under Salvador E. Luria on bacteriophages, the viruses that invade bacteria in order to reproduce—a topic for which Luria received a Nobel Prize in medicine in 1969. Watson then went to Denmark for postdoctoral work—to continue studying viruses and to remedy his relative ignorance of chemistry. At a conference at the

■ Rosalind Franklin in Paris.
Courtesy Vittorio Luzzati.

■ James Watson and Francis Crick with their DNA model at the Cavendish Laboratories in 1953. Photograph by C. Barrington Brown. Courtesy James Dewey Watson.

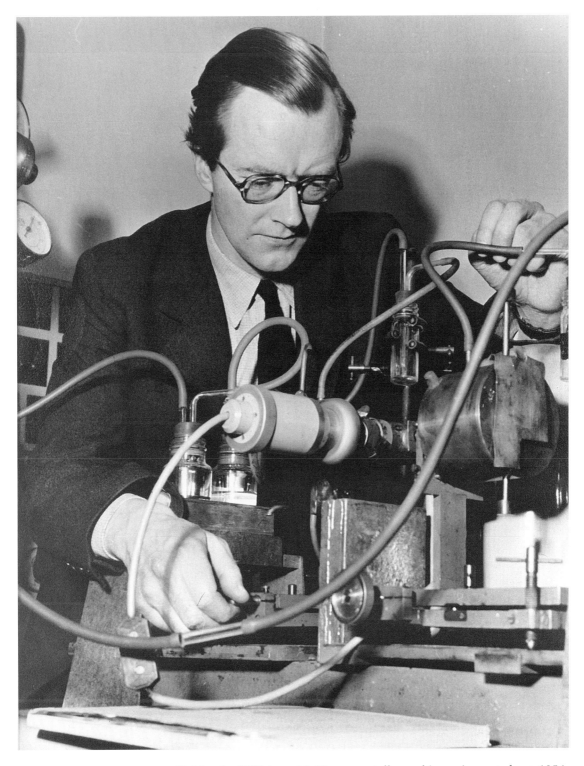

■ Maurice Wilkins with X-ray crystallographic equipment about 1954. Courtesy King's College, London, and Horace Freeland Judson.

Zoological Station at Naples, Watson heard Wilkins talk on the molecular structure of DNA and saw his recent X-ray crystallographic photographs of DNA—and was hooked.

Watson soon moved to the Cavendish Laboratory. There several important X-ray crystallographic projects were in progress under William Lawrence Bragg's leadership (see Bragg, earlier in this chapter), including Max Perutz's investigation of hemoglobin and John Kendrew's study of myoglobin—a protein in muscle tissue that stores oxygen. (Perutz and Kendrew received Nobel Prizes in chemistry for their work in the same year that the prizes were awarded to the DNA researchers—1962.) Working under Perutz was Francis Crick, who had earned a bachelor's degree in physics from University College, London, and had helped develop radar and magnetic mines during World War II. Crick, another physicist in biology, was supposed to be writing a dissertation on the X-ray crystallography of hemoglobin when Watson arrived, eager to recruit a colleague for work on DNA. Inspired by Pauling's success in working with molecular models, Watson and Crick rapidly put together several models of DNA and attempted to incorporate all the evidence they could gather. Franklin's excellent X-ray photographs, to which they had gained access without her permission, were critical to the correct solution. The four scientists announced the structure of DNA in articles that appeared together in the same issue of *Nature*.

Then they moved off in different directions. Franklin went to Birkbeck College, London, to work in J. D. Bernal's laboratory—a much more congenial setting for her than King's College. Before her death she made important contributions to the X-ray crystallographic analysis of the structure of the tobacco mosaic virus—a landmark in the field. Wilkins applied X-ray techniques to the structural determination of nerve cell membranes and of ribonucleic acid (RNA)—a molecule that is associated with chemical synthesis in the living cell—while rising in rank and responsibility at King's College. Watson's subsequent career eventually took him to Cold Spring Harbor Laboratory of Quantitative Biology on Long Island, where as director from 1968 onward he led it to new heights as a center of research in molecular biology. From 1988 to 1992 he headed the National Center for Human Genome Research at the National Institutes of Health. Crick stayed at Cambridge and made fundamental contributions to unlocking the genetic code. He and Sydney Brenner demonstrated that each group of three adjacent bases on a single DNA strand codes for one specific amino acid. He also correctly hypothesized the existence of "transfer" RNA, which mediates between "messenger" RNA and amino acids. After twenty years at Cambridge, with several visiting professorships in the United States, Crick joined the Salk Institute for Biological Studies in La Jolla, California.

8. Petroleum and Petrochemicals

This chapter celebrates scientists who achieved critical breakthroughs in refining oil and natural gas to produce gasoline and other substances used for fuel and lighting, as well as a wide variety of other chemicals. Catalysts form a key element of this story. They can reduce the pressures and temperatures needed for carrying out commercially useful reactions and ensure the specificity of products—thus reducing costs and potentially harmful waste products.

GEORGE O. CURME (1888–1976)

George Curme is known for his pioneering work on ways to use by-products of petroleum processing. After completing his doctorate in chemistry at the University of Chicago, Curme went to Germany to study with Fritz Haber (see Haber, Chapter 2) and Emil Fischer, who had already won the Nobel Prize in chemistry (1902) for his syntheses of sugars, purine derivatives, and peptides. Curme remained in Germany until the outbreak of World War I sent him back to the United States. He found a job with the Mellon Institute in Pittsburgh, where the Prest-O-Lite Company had established a fellowship for research on new ways to manufacture acetylene. The company used acetylene in bicycle and automobile lamps as well as in oxyacetylene torches. Curme succeeded in producing acetylene from petroleum with a high-frequency electric arc, but the process also produced a substantial amount of ethylene, for which there were no uses. Curme then began to investigate ways to produce other chemicals from ethylene. His first breakthrough came during World War I, when he tried to make mustard gas from ethylene, but instead synthesized ethylene glycol, which was eventually used in antifreeze.

In 1917, when Prest-O-Lite was absorbed during the formation of the Union Carbide and Carbon Corporation, Curme focused his efforts on cracking ethane to produce ethylene. Encouraged by the progress of this research, in 1920 the company set up the Carbide and Carbon Chemical Company and bought a gasoline refinery in Clendenin, West Virginia, to supply the necessary ethane and other light hydrocarbon gases—until then unused by-products. Over the next decade the new company created ethylene glycol antifreeze, synthetic ethyl alcohol, and dozens of other industrial chemicals. The Union Carbide venture into petrochemicals was a trendsetter for the whole chemical industry.

■ George O. Curme in his Mellon Institute laboratory before World War I, where he laid the foundation for the petrochemicals revolution that took place after World War II. Courtesy Union Carbide Corporation.

■ George Curme and colleagues standing outside the laboratory called "The Shack" at the Mellon Institute. Curme is standing second from the left; his brother Henry R. Curme, wearing a laboratory coat, stands next to him. Henry was killed in a chemical explosion. Courtesy Robert C. Hieronymus.

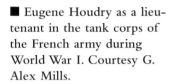

■ Eugene Houdry as a lieutenant in the tank corps of the French army during World War I. Courtesy G. Alex Mills.

EUGENE HOUDRY (1892–1962)

One of the first improvements in petrochemical production was the process developed by Eugene Houdry for "cracking" petroleum molecules into the shorter ones that constitute gasoline. (Earlier commercial processes for cracking petroleum relied instead on heat.)

Eugene Houdry obtained a degree in mechanical engineering in his native France before joining the family metalworking business in 1911. After he served in the tank corps in World War I—for which he received honors for extraordinary heroism in battle—he pursued his interest in automobiles (especially race cars) and their engines. On a trip to the United States he visited the Ford Motor Company factory and attended the Indianapolis 500 race. His interest soon narrowed to improved fuels. Because France produced little petroleum—and the world supply was thought to have nearly run out—Houdry, like many other chemists and engineers, searched for a method to make gasoline from France's plentiful lignite (brown coal). After testing hundreds of catalysts to effect the hoped-for molecular rearrangement, Houdry began working with silica-alumina and changed his feedstock from lignite to heavy liquid tars. By 1930 he had produced small samples of gasoline that showed promise as a motor fuel.

In the early 1930s Houdry collaborated with two American oil companies, Socony Vacuum and Sun Oil, to build pilot plants. Oil companies that did not want to resort to the new additive tetraethyl lead were eagerly looking for other means to increase octane

levels in gasoline. In 1937 Sun Oil opened a full-scale Houdry unit at its refinery in Marcus Hook, Pennsylvania, to produce high-octane Nu-Blue Sunoco gasoline. By 1942, fourteen Houdry fixed-bed catalytic units were bearing the unanticipated burden of producing high-octane aviation gasoline for the armed forces.

(One limitation of the process was that it deposited coke on the catalyst, which required that the unit be shut down while the coke was burned off in a regeneration cycle. Warren K. Lewis and Edwin R. Gilliland [see Lewis, Chapter 10] of the Massachusetts Institute of Technology, who were hired as consultants to Standard Oil Company of New Jersey [now Exxon], finally solved this problem with great ingenuity and effort. They developed the "moving bed" catalytic converter, in which the catalyst was itself circulated between two enormous vessels, the reactor and the regenerator.)

Houdry continued his work with catalysts and became particularly fascinated with the catalytic role of enzymes in the human body and the changes in enzyme-assisted processes caused by cancer. About 1950, when the results of early studies of smog in Los Angeles were published, Houdry became concerned about the role of automobile exhaust in air pollution and founded a special company, Oxy-Catalyst, to develop catalytic converters for gasoline engines—an idea ahead of its time. But until lead could be eliminated from gasoline (lead was introduced in the 1920s to raise octane levels), it poisoned any catalyst.

■ Eugene Houdry in 1953, holding a small catalytic converter. Courtesy Sun Company.

JOHN H. SINFELT (1931–)

John Sinfelt attended a two-room schoolhouse in a small village twenty-five miles from Pennsylvania State University, then attended the university, majoring in chemical engineering, and proceeded to the University of Illinois for a doctorate. He started research at Standard Oil Development Company (now Exxon Research and Engineering) in 1954 on improving the platinum catalysts that Vladimir Haensel had pioneered at the Universal Oil Products Company for the production of gasoline. Sinfelt believed that the overlapping chemistry and chemical engineering curricula at Pennsylvania State University and the University of Illinois had prepared him well for this assignment. After developing a new approach to bimetallic catalysts, which he called "clusters," Sinfelt invented a superior platinum-iridium catalyst that was important in the quest to produce lead-free, high-octane gasolines cheaply. His work also provided a scientific base for other processes to produce petrochemicals in high volume.

■ John H. Sinfelt in his laboratory at Exxon Research and Engineering Company. Courtesy Exxon Research and Engineering Company.

■ Paul B. Weisz with a model that shows how tightly a paraffin molecule fits in a channel of a zeolyte. The catalyst cracks the paraffin into smaller gas molecules. Photo by André Grossman. Reproduced with permission.

PAUL B. WEISZ (1919–)

While working at Mobil Oil, Paul Weisz pioneered the use of natural and synthetic zeolites (hydrous silicates) as catalysts. These catalysts are highly selective, facilitating only certain reactions between specific molecules of given shapes. Processes based on zeolite catalysts were first developed in the 1960s and were found to increase both the amount of gasoline obtainable from petroleum and the octane rating of gasoline. Shape-selective zeolite catalysts proved to be widely applicable to many other industrial processes, including the manufacture of gasoline from natural gas and the production of raw materials for making polyester garments, plastics, and other products from petroleum.

Weisz, who grew up in Berlin, knew from an early age that he wanted to be a scientist and go to America. As a radio "ham," who at the age of sixteen had published articles in a radio journal, Weisz managed to get a summer job with Telefunken, a major radio and electronics company; he later interned at the Cosmic Ray Institute after being turned down by Hahn and Meitner's institute because those scientists were working on "classified" research (see Hahn and Meitner, Chapter 5).

125

Meanwhile, his goal of coming to the United States was still out of reach—Weisz had no American relatives and Germany banned the export of monetary funds. But at age sixteen Weisz wrote to three American universities offering his family's support of an American wishing to study in Germany in exchange for similar support for him in the United States. Such an exchange was worked out through Auburn University (formerly Alabama Technical University). He arrived in Alabama in 1939, narrowly escaping the outbreak of war. Having interrupted his graduate work in Berlin, Weisz completed the credits for a bachelor's degree in less than one year. He then continued research on cosmic radiation at the Bartol Research Foundation of the Franklin Institute in Swarthmore, Pennsylvania. After the United States entered World War II, Weisz, a longtime radio enthusiast, became an electronics engineer, teaching Signal Corps trainees first at Swarthmore College and later at the Radiation Laboratory at the Massachusetts Institute of Technology, where they were participating in the development of LORAN, or long-range aid to navigation, based on radio signals.

Not until after he had become a well-established researcher in catalytic chemistry did Weisz achieve his long-deferred goal of a doctoral degree at the Eidgenössische Technische Hochschule in Zürich, Switzerland, while on a leave of absence from Mobil. His thesis on the mechanism of dyeing fibers developed some of the basic laws about the entrance of dyes into fibers, based on his experience with the velocity with which chemicals flow into catalytic materials.

After retiring from Mobil, Weisz began yet another career—applying general chemical and physical principles to biomedical research. He designed (and Madeleine Joullié and one of her students at the University of Pennsylvania synthesized) molecules that mimic some of the healing properties of heparin but that do not exhibit heparin's potentially dangerous anticoagulant effects. Acting in some ways like an *organic* zeolite, these molecules can capture other molecules, such as cortisone. The combination of cortisone and the new molecule inhibits the proliferation of blood capillaries associated with tumor growth and some forms of blindness. Other applications are under investigation.

9. Plastics
and
Other Polymers

Understanding how to make small molecules polymerize into long-chain hydrocarbons to form plastics and other synthetics was closely related to the efforts of chemists and engineers to break down and utilize fragments of the long-chain hydrocarbons in petroleum. Scientists active in the area of petrochemicals have shared their knowledge of catalysts and increased the variety and quantity of chemicals usable in polymerization processes that create all sorts of synthetic materials.

Leo Hendrik Baekeland (1863–1944)

One of the earliest synthetics that transformed the material basis of modern life was Bakelite, a polymeric plastic made from phenol and formaldehyde. Leo Hendrik Baekeland invented Bakelite in 1907, and his inventive and entrepreneurial genius also propelled him into several other new chemical technological ventures at the turn of the twentieth century.

After completing his doctorate at the University of Ghent in his native Belgium, Baekeland taught for several years. In 1889, when he was twenty-six, he traveled to New York on a fellowship (that had also allowed him to visit universities in England, Scotland, and Germany) to continue his study of chemistry. Professor Charles F. Chandler (see Chandler, Chapter 11) of Columbia University then persuaded Baekeland to stay in the United States and recommended him for a position at a New York photographic supply house. This experience led him a few years later, when he was working as an independent consultant, to invent Velox, an improved photographic paper that could be developed in gaslight rather than sunlight. In 1898 the Eastman Kodak Company purchased Baekeland's invention for a reputed $750,000, a sum that allowed him to spend the rest of his life in experimentation.

Baekeland next entered the field of electrochemistry. He visited Berlin briefly to update his knowledge of this new area of study, and he equipped his private laboratory on the grounds of his home in Yonkers, New York, with a few electrochemical appliances. At the request of Elon Hooker, Baekeland cooperated with Clinton P. Townsend, the inventor of a new electrolytic cell for producing caustic soda and chlorine from salt, in setting up a pilot plant at the Brooklyn Edison Station. The success of their experiment led Elon Hooker to form Hooker Electrochemical Company in Niagara Falls—now part of the Oxychem subsidiary of Occidental Petroleum.

When friends asked Baekeland how he entered the field of synthetic resins, he answered that he had chosen it deliberately, looking for a way to make money. His

■ *Above:* Leo Hendrik Baekeland. Courtesy Edgar Fahs Smith Memorial Collection, Department of Special Collections, University of Pennsylvania Library.
Right: The original Bakelizer, used by Baekeland and his coworkers from 1907 to 1910 to form Bakelite by reacting phenol and formaldehyde under pressure at high temperature. Courtesy Smithsonian Institution.

■ Leo Hendrik Baekeland with his wife Céline and their children, Nina and George, on a family outing at Snug Rock, Yonkers, New York, around 1900. Like several other chemical industry pioneers, Baekeland had a love affair with the automobile— which turned out to be an engine of progress in chemicals manufacture in the United States. Courtesy Smithsonian Institution.

first objective was to find a replacement for shellac, which at that time was made from the shells of oriental lac beetles. Chemists had begun to recognize that many of the natural resins and fibers useful for coatings, adhesives, woven fabrics, and the like were polymers (large molecules made up of repeating structural units), and they had begun to search for combinations of reagents that would react to form *synthetic* polymers. Baekeland began to investigate the reactions of phenol and formaldehyde, and first produced a soluble phenol-formaldehyde shellac called "Novolak," which never became a market success. Then he turned to developing a binder for asbestos, which at that time was molded with hard natural rubber. By carefully controlling the pressure and temperature applied to an intermediate made from the two reagents, he could produce a polymer that, when mixed with fillers, produced a hard moldable plastic. Bakelite, though relatively expensive, was soon found to have many uses, especially in the rapidly growing automobile and radio industries. Baekeland retired in 1939 to sail his yacht, the *Ion,* among other activities, and sold his successful plastics company to the Union Carbide and Carbon Corporation.

E. I. DU PONT DE NEMOURS & COMPANY
CHEMICAL DEPARTMENT

■ Wallace Hume Carothers in a characteristically pensive mood, in the early 1930s. Courtesy Hagley Museum and Library.

WALLACE HUME CAROTHERS (1896–1937)

Two of the twentieth century's most widely used synthetic polymers—neoprene and nylon—originated in 1930 in the research laboratory of Wallace Hume Carothers at the DuPont Company.

After high school Carothers attended Capital City Commercial College in Des Moines, Iowa, in a program of accountancy and secretarial administration. He then went on to a four-year college, Tarkio College in Missouri, to complete a bachelor's degree in chemistry. After a year of teaching at the University of South Dakota, he proceeded to the University of Illinois, where he earned his doctorate in 1924. As a young instructor at Harvard University, Carothers was already pursuing research in polymers when DuPont's Charles Stine recruited him for the fundamental research program that Stine was then organizing. Elmer K. Bolton, Carothers's immediate boss, asked him to investigate the chemistry of an acetylene polymer that might lead to a synthetic rubber. In April 1930 one of Carothers's assistants, Arnold M. Collins, isolated a new liquid compound, chloroprene, which spontaneously polymerized to produce a rubberlike solid. The new polymer was similar chemically to natural rubber, which encouraged Bolton to exploit it. Neoprene, as the product was named, was superior to the natural product in some applications and became the first commercially successful, though never inexpensive, specialty rubber.

Soon after this discovery another Carothers associate, Julian W. Hill, created a strong, elastic synthetic fiber while trying to produce superpolymers of high molecular weight by reacting glycols and dibasic acids with strong acids under reduced pressure in a molecular still. The resultant early *polyesters*, however, were problematic: They had such low melting points and high water solubility that they were not commercially viable. After a few attempts to solve these problems, Carothers discontinued this line of research. But Bolton encouraged him not to give up on the wider field of fibers. When Carothers finally renewed work in that

■ Wallace Carothers, in a happier mood, holding Louisa Hill—the daughter of Julian Hill, who carried out the original polymerization that led to nylon. Courtesy Hagley Museum and Library.

area in early 1934, he and his team used amines rather than glycols to produce *polyamides* rather than polyesters. Polyamides are synthetic proteins and are more stable than polyesters, which are structurally similar to natural fats and oils.

Carothers's group soon discovered an outstanding polyamide fiber. Bolton played a key role in the development of the discovery, later named "nylon." In the years that followed, Carothers's scientific creativity was

■ Giant leg, thirty-five feet high, advertises nylons to the Los Angeles area. Courtesy Hagley Museum and Library.

crippled by worsening bouts of depression that finally prompted his suicide in April 1937, just when the true magnitude of the discovery of nylon was becoming apparent. By this time Bolton had decided to commercialize nylon by creating an alternative product to the lucrative silk stocking market, leaving other applications for later. Nylon went into production in 1939,

and the display of the new stockings was a sensation at the World's Fair in New York that year. With the onset of World War II, nylon was commandeered for war purposes—for example, to make parachute canopies. Once the war was over, sales to civilian consumers skyrocketed.

ROY J. PLUNKETT (1910–1994)

From the 1930s to the present, beginning with neoprene and nylon, the American chemical industry has introduced a cornucopia of polymers to the consumer. Teflon, discovered by Roy Plunkett at DuPont's Jackson Laboratory in 1938, was an accidental invention—unlike most of the other polymer products. But as Plunkett often told student audiences, his mind was prepared by education and training to recognize novelty.

As a poor Ohio farm boy during the Depression, Plunkett attended Manchester College, operated by the Church of the Brethren. His roommate for a time at this small college was Paul Flory, who would win the 1984 Nobel Prize in chemistry for his contributions to the theory of polymers. Like Flory, Plunkett went on to Ohio State University for a doctorate, and also like Flory he was hired by DuPont. Unlike Flory, Plunkett made his entire career at DuPont.

Plunkett's first assignment at DuPont was researching new chlorofluorocarbon refrigerants—then seen as great advances over earlier refrigerants like sulfur dioxide and ammonia, which regularly poisoned food-industry workers and people in their homes. Plunkett had produced one hundred pounds of tetrafluoroethylene gas (TFE) and stored it in small cylinders at dry-ice temperatures preparatory to chlorinating it. When he and his helper prepared a cylinder for use, none of the gas came out—yet the cylinder weighed the same as before. They opened it and found a white powder, which Plunkett had the presence of mind to characterize for properties other than refrigeration potential. He found the substance to be heat resistant and chemically inert, and to have very low surface friction so that most other substances would not adhere to it. Plunkett realized that against the predictions of polymer science of the day, TFE had polymerized to produce this substance—later named "Teflon"—with such potentially useful characteristics. Chemists and engineers in the Central Research Department who had special experience in polymer research and development investigated the substance further. Meanwhile, Plunkett was transferred to the tetraethyl lead division of DuPont, which produced the additive that for many years boosted gasoline octane levels.

At first it seemed that Teflon was so expensive to produce that it would never find a market. Its first use was fulfilling the requirements of the gaseous diffusion process of the Manhattan Project for materials that could resist corrosion by fluorine or its compounds (see Landau, Chapter 10). Teflon pots and pans were invented years later. The awarding of Philadelphia's Scott Medal in 1960 to Plunkett—the first of many honors for his discovery—provided the occasion for the introduction of Teflon bakeware to the public: Each guest at the banquet went home with a Teflon-coated muffin tin.

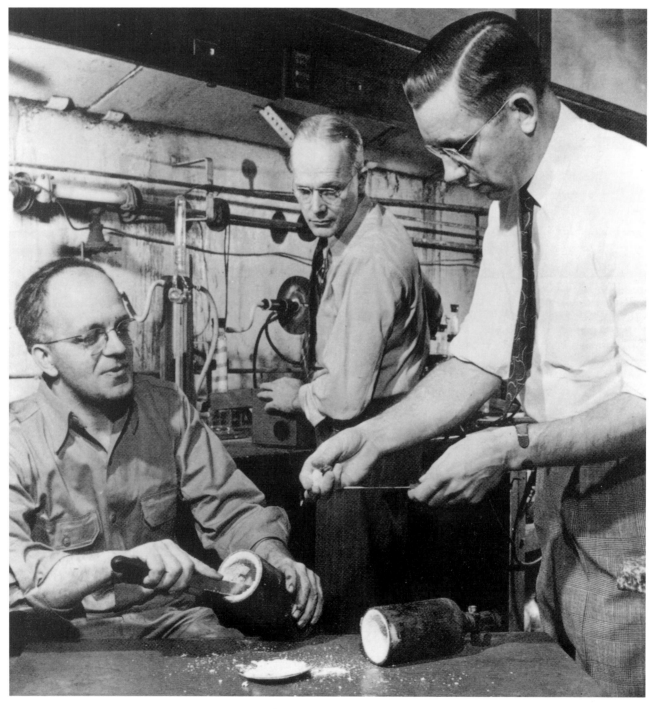

■ Re-enactment of the 1938 discovery of Teflon. Left to right, Jack Rebok, Robert McHarness, and Roy Plunkett. Courtesy Hagley Museum and Library.

■ Roy Plunkett with a cable insulated with Teflon and a Teflon-coated muffin tin. Gift of Roy Plunkett.

WALTER LINCOLN HAWKINS (1911–1992)

For years the communications industry has been one of the largest users of polymer products, especially for wire and cable coatings. But exposure to sunlight and extreme temperatures causes these coatings to deteriorate, a problem that Walter Lincoln Hawkins and Vincent L. Lanza of Bell Laboratories tackled in 1956. They first developed a special plastics additive that inhibits oxidation of the polymeric material. Hawkins then designed a laboratory test using infrared spectroscopy to predict the weatherability of a plastic surface, working also with Maureen G. Chan.

A native of Washington, D.C., Hawkins was orphaned at an early age and raised by a sister. Yet in the middle of the Depression he found enough family support and personal confidence to disregard the advice of well-meaning teachers who were convinced that an African American could never become a chemical engineer. Hawkins headed to Rensselaer Polytechnic Institute to complete his engineering degree and then went on to receive a master's degree in chemistry at Howard University and a doctorate at McGill University in Montreal, where he specialized in cellulose chemistry. After a postdoctoral fellowship at Columbia University he was hired in 1942 by Bell Laboratories, where he was the first African-American scientist on the staff. He eventually became head of plastics chemistry research and development and assistant director of the Chemical Research Laboratory.

Hawkins was an active leader in efforts to expand the nation's pool of scientific talent, as first chairman of the American Chemical Society's Project SEED (Summer Educational Experience for the Economically Disadvantaged)—created to promote chemistry among economically disadvantaged youth—and as a board member of several educational institutions. He was honored on several occasions for this work and for his contributions to polymer science, including the award of the National Medal of Technology, presented to him by the President of the United States just two months before his death.

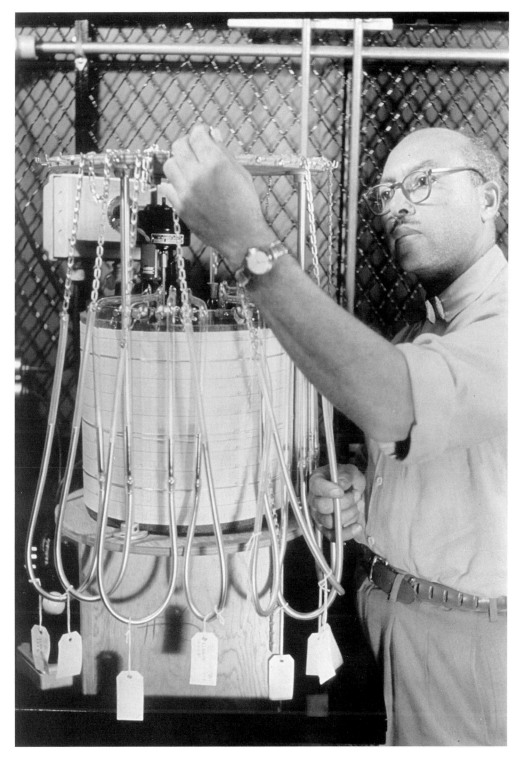

■ W. Lincoln Hawkins of Bell Laboratories conducting accelerated oxidation tests on polyethylene to determine the effectiveness of various antioxidants. Courtesy AT & T Archives.

HENRY AARON HILL (1915–1979)

The growing polymer products industry had created a demand for the chemical intermediates used in various polymerization processes, so chemical and petrochemical companies large and small turned to producing these substances. National Polychemicals, Inc., founded in 1952 by Henry Aaron Hill, was just such a supplier of intermediates.

Born in St. Joseph, Missouri, Hill graduated with a bachelor's degree in chemistry from Johnson C. Smith University in North Carolina. After a year of graduate study at the University of Chicago, Hill went on to the Massachusetts Institute of Technology, where he received a doctorate in organic chemistry in 1942. At MIT he met Professor James Flack Norris, a pioneer physical organic chemist and former president of the American Chemical Society, "the first big man . . . who was more interested in my ability to learn chemistry than in the identity of my grandparents." As an African American, Hill often encountered prejudice, the prob-able reason that he had to send out fifty-four applications before he landed a job with North Atlantic Research Corporation of Newtonville, Massachusetts. He eventually rose to be vice president while doing research and development on water-based paints, fire-fighting foam, and several types of synthetic rubber. After leaving North Atlantic Research he worked as a group leader in the research laboratories of Dewey and Almy Chemical Company before starting his own entrepreneurial venture—National Polychemicals. Ten years later he founded Riverside Research Laboratories, which offered research, development, and consulting services in polymer production.

From early in his career Hill was active in the American Chemical Society—most memorably in establishing standards for employer–employee relationships in the chemical profession and as the society's first African-American president (1977).

■ Henry Aaron Hill standing outside the American Chemical Society building in Washington, D.C. Courtesy American Chemical Society.

■ Stephanie Kwolek at a polarizing microscope.
Gift of Stephanie Kwolek.

STEPHANIE L. KWOLEK (1923–)

In 1965 Stephanie Kwolek succeeded in creating the first of a family of synthetic fibers of exceptional strength and stiffness. The best known member is Kevlar, a material used in bulletproof vests as well as in boats, airplanes, ropes, cables, tires, tennis racquets, skis, and so forth—in total about two hundred applications.

Kwolek was born in New Kensington, Pennsylvania. Her father, who died when she was ten years old, was a naturalist by avocation. She spent many hours with him exploring the woods and fields near her home and filling scrapbooks with leaves, wildflowers, seeds, grasses, and pertinent descriptions. From her mother, first a homemaker and then by necessity a career woman, Kwolek inherited a love of fabrics and sewing. At one time she thought she might become a fashion designer, but her mother warned her that she would probably starve in that business because she was such a perfectionist. Later Kwolek became interested in teaching and then in chemistry and medicine.

When she graduated from the women's college (Margaret Morrison Carnegie College) of Carnegie-Mellon University, she applied for a position as a chemist with the DuPont Company, among other places. Her job interview with W. Hale Charch, who had invented the process to make cellophane waterproof and who was by then a research director, was a memorable one. After Charch indicated that he would let her know in about two weeks whether she would be offered a job, Kwolek asked him if he could possibly make a decision sooner since she had to reply shortly to another offer. Charch called in his secretary and in Kwolek's presence dictated a job offer letter. In later years, reflecting upon this bold request for a woman to make in 1946, she suspected that her assertiveness influenced his decision in her favor. At DuPont the polymer research she worked on was so interesting and challenging that she decided to drop her plans for medical school and make chemistry a lifetime career.

She was engaged in several projects, including a search for new polymers as well as new condensation processes that take place at lower temperatures—about 0° to 40° C. The familiar melt condensation

■ *Left:* Bulletproof vest and helmet employing Kevlar. Courtesy DuPont.
Right: Kevlar is widely used in sports equipment. Courtesy DuPont.

■ Stephanie L. Kwolek and others of the group that developed Kevlar. Left to right, Kwolek, Herbert Blades, Paul W. Morgan, and Joseph L. Rivers, Jr. Gift of Stephanie Kwolek.

polymerization process used in preparing nylon, for example, was instead done at more than 200° C. The lower-temperature polycondensation processes, which employ very fast-reacting intermediates, make it possible to prepare polymers that are thermally unstable or cannot be melted.

Kwolek was in her forties when she was asked to scout for the next generation of high-performance fibers. This assignment involved preparing intermediates, synthesizing paraoriented aromatic polyamides of high molecular weight, dissolving the polyamides in solvents, and spinning these solutions into fibers. She unexpectedly discovered that under certain conditions large numbers of the molecules of these rod-like polyamides become lined up in parallel, that is, form liquid crystalline solutions, and that these solutions can be spun directly into oriented fibers of very high strength and stiffness. These polyamide solutions were unlike any polymer solutions previously prepared in the laboratory. They were unusually fluid, turbid, and buttermilk-like in appearance, and became opalescent when stirred. The person in charge of the spinning equipment initially refused to spin the first such solution because he feared that the turbidity was caused by the presence of par-

ticles that would plug the tiny holes (0.001 inch in diameter) in the spinneret. He was finally persuaded to spin, and much to his surprise, strong, stiff fibers were obtained with no difficulty. Following this breakthrough many fibers were spun from liquid crystalline solutions, including the yellow Kevlar fiber.

Kwolek has received many awards for her invention of the technology behind Kevlar fiber, including induction into the National Inventors Hall of Fame in 1994 as only the fourth woman member of 113. In 1996 she received the National Medal of Technology, and in 1997 the Perkin Medal, presented by the American Section of the Society of Chemical Industry—both honors rarely awarded to women. She has served as a mentor for other women scientists and participated in programs that introduce young children to science. One of Kwolek's most cited papers, written with Paul W. Morgan, is "The Nylon Rope Trick" (*Journal of Chemical Education*, April 1959, 36:182–184). It describes how to demonstrate condensation polymerization in a beaker at atmospheric pressure and room temperature—a demonstration now common in classrooms across the nation.

10. Chemical Engineering: Chemistry Scales Up for Industry

All of the chemical products and processes created by the chemical achievers described in this work had to be scaled up from the laboratory to the industrial plant to give them commercial importance. Scaling up was traditionally carried out by chemists—often *industrial* chemists specially trained in the chemistry of industrial processes—working with mechanical engineers. The notion of a new kind of engineer, one who understood both chemical processes and mechanical equipment, was first broached in England around 1880. The idea first took firm root, however, in the United States in the 1890s.

GEORGE E. DAVIS (1850–1906)

George Davis is generally credited with initiating the concept of chemical engineering. Davis studied at the Slough Mechanics Institute and the Royal School of Mines in London (now part of Imperial College, London) and then headed north to work in the chemical industry around Manchester. Before he embarked on a career as a consulting engineer, he held various positions—one as an inspector for the Alkali Act of 1863, a very early piece of environmental legislation that required soda manufacturers to reduce the amount of hydrochloric acid gas vented into the atmosphere from their factories. Davis was also a moving spirit behind the formation of the Society of Chemical Industry

(1881), which he had wanted to name the Society of Chemical Engineering.

In 1887 Davis gave a series of twelve lectures at the Manchester School of Technology, which formed the basis of his *Handbook of Chemical Engineering* (1901; revised 1904)—the first of its kind. There were already industrial chemistry books written for each chemical industry—for example, alkali manufacture, acid production, brewing, and dyeing—but Davis organized his text by the basic operations common to many industries—transporting solids, liquids, and gases; distillation; crystallization; and evaporation, to name a few.

■ George E. Davis.

Arthur D. Little (1863–1935), William H. Walker (1869–1934), and Warren K. Lewis (1882–1975)

According to Davis, the ideal chemical engineer could move from industry to industry, mixing and matching the various operations that the American Arthur D. Little was later to call "unit operations." Little first used this term in a 1915 report to the president of the Massachusetts Institute of Technology, where the curriculum in "chemical engineering" dated from 1888. The first degrees in this field in the United States were given at MIT in 1891, although the content of the original courses still centered on industrial chemistry and mechanical engineering, without the characteristic unit-operations laboratory. William H. Walker, Warren K. Lewis, and Little were among the leaders who defined chemical engineering as a separate profession with a distinct approach and training method.

Little majored in chemistry at MIT before the advent of chemical engineering and was the editor of the college newspaper, *The Tech*—an experience that prepared him for his role as a spokesperson for chemical engineering education, industrial research, and the American chemical industry. His first jobs made him an expert in the new sulfite process for making paper, and in 1886 he and a coworker, Roger B. Griffin, set up a consulting company whose successor, Arthur D. Little, Inc., prospers to this day. Seven years after they founded this company, Griffin died tragically in a laboratory explosion, but Little continued as a consultant, becoming involved in two new synthetics, cellulose nitrate and cellulose acetate, which were being used for photographic film and woven fabrics. The reluctance of American financiers to undertake ventures in this new technology—opportunities that were instead seized by Europeans starting up plants in the United States—prompted Little to mount a writing and speaking campaign directed at financial, political, and educational leaders to encourage the nascent American chemical industry. From an early date Little also preached against heedless industrial practices, referring ominously to "the

handwriting on the wall" for a society that would destroy its own environment.

In 1900 Little formed a new partnership with William H. Walker, a young MIT chemistry instructor who had graduated from Pennsylvania State University with a bachelor's degree in chemistry and who held a Ph.D. in organic chemistry from the University of Göttingen. Walker was soon recalled to MIT to reform the chemical engineering curriculum—with Little in the background gaining approval and funds for various initiatives to bring industry and education closer together. In 1908 MIT's Research Laboratory of Applied Chemistry commenced operations. Here chemical engineering students worked on real industrial problems supplied by various chemical companies, who also supplied fellowships.

In 1908 Warren K. Lewis, a young graduate of MIT's chemical engineering program and a Ph.D. in organic chemistry from the University of Breslau, joined the teaching staff at MIT. He contributed a great deal to the program with his ability to view engineering problems theoretically and mathematically and with his memorable teaching style, which was sometimes described as bombastic but nonetheless endeared him to students. In 1916 three plant-based stations of the School of Chemical Engineering Practice were inaugurated, thus enabling students to gain more hands-on experience by spending eight weeks at one of the stations under the supervision of an MIT faculty member. Meanwhile, under Lewis's guidance, teaching in the fundamental unit-operations course had become highly quantitative. One result of this course, *The Principles of Chemical Engineering* (1923), by Walker, Lewis, and William H. McAdams, became the standard text for chemical engineering instruction for decades.

Shortly after publication of the book, Walker returned to consulting. From his faculty position Lewis also maintained close connections with industry, consulting most

145

■ Arthur D. Little in 1922. Courtesy MIT Museum.

frequently for Standard Oil Company of New Jersey (now Exxon). His most famous industrial contribution—in collaboration with his colleague, Edwin R. Gilliland—was fluidized-bed catalytic cracking of petroleum (see Houdry, Chapter 8), which solved many of the problems posed by the Houdry process and played a large part in supplying the vast quantities of gasoline needed during World War II.

■ Warren K. Lewis teaching at MIT's School of Chemical Engineering Practice at the Bayway Refinery of Standard Oil Company of New Jersey. Courtesy Exxon Corporation.

■ William H. Walker with slide rule at the ready. Courtesy MIT Museum.

DONALD OTHMER (1900–1995)

Donald Othmer, co-editor of the *Kirk-Othmer Encyclopedia of Chemical Technology*, held more than 150 U.S. and foreign patents, most of which he obtained while working as a full-time professor of chemical engineering at Polytechnic Institute of New York (formerly Brooklyn Polytechnic).

Othmer's high school teachers in solid geometry and chemistry alerted him to the then new field of chemical engineering. From Central High School in Omaha, Nebraska, he went to the Armour Institute in Chicago (now Illinois Institute of Technology), the University of Nebraska, and the University of Michigan. With a Michigan doctorate in chemical engineering in hand, he went to work for Eastman Kodak in Rochester, New York, which was then converting to the production of "safety" film—cellulose acetate—instead of the dangerously explosive cellulose nitrate. While working out the problems of acetic acid recovery from the acetate-making process, Othmer invented a basic laboratory device, the "Othmer still," for the simple and precise determination of vapor–liquid equilibrium data. It is used in industry worldwide and is also commonly found in physical chemistry teaching laboratories. When the Depression forced a slowdown in Kodak's expansion, Othmer decided to run his own business as an independent consultant, until lean economic times made selling patent rights an increasingly problematic activity.

In 1932 Othmer joined the Brooklyn Polytechnic faculty, a position that provided him economic stability. There he collaborated on the *Encyclopedia of Chemical Technology* with Raymond Kirk, a colleague in the chemistry department. Othmer continued to devise process innovations and patent them, commonly following up topics introduced by his students, who were often already working in the chemical industry but taking classes in the evenings or weekends to complete academic degrees. Othmer's patents cover methods, processes, and equipment for the manufacture and pro-

■ Donald Othmer as a young man.

cessing of chemicals, solvents, synthetic fibers, acetic acid, and methanol. He long advocated using methanol as a fuel for motor vehicles because it contains fewer pollutants and in the long run is more plentiful than gasoline. Other projects included providing potable water through desalinization plants and improving the processing of domestic and industrial sewage. Vacations and summers found Othmer traveling worldwide to advise on the installation of his various patented processes.

Donald Othmer and his wife Mildred donated generously over the years to support medical care facilities and institutions devoted to chemistry and chemical engineering.

■ Donald Othmer, with a gold-plated Othmer still in the background.

■ Ralph Landau (second from left) welcoming Prince Bernhard of the Netherlands (right) to the Rotterdam plant of Oxirane, a joint venture of Scientific Design, Inc., and Arco Chemical Company. The burgomaster of Rotterdam, wearing a hardhat, stands at the left, and the plant manager faces the camera. Courtesy Ralph Landau.

RALPH LANDAU (1916–)

Ralph Landau founded one of the world's most successful engineering and design firms, Scientific Design, Inc., and its successor corporations, which developed and commercialized nearly a dozen processes for producing petrochemicals.

As a high school student in Philadelphia during the Depression, Landau noticed a newspaper article about the new "glamour field" of chemical engineering and decided that it was the career for him. A scholarship student at the University of Pennsylvania, he majored in chemical engineering and went on to receive his doctorate from Massachusetts Institute of Technology. As a new Ph.D. he worked for M. W. Kellogg Company, one of the first engineering firms that specialized in design and development for the oil refining and

chemical industries. During World War II, Kellogg was asked to build a large-scale facility at Oak Ridge, Tennessee, to separate uranium 235, which was needed for the atomic bomb, from the predominant isotope, uranium 238. Landau was given the responsibility of designing the equipment to produce fluorine, a highly reactive substance needed to make the uranium hexafluoride used in the gaseous diffusion process. He also oversaw the production of the fluorinated compounds used to protect surfaces in contact with the uranium hexafluoride.

After the war he and a construction engineer he had met at Oak Ridge, Harry Rehnberg, started Scientific Design with the objective of improving petrochemical production processes. Much of their business in the

■ Ralph Landau. Photograph by Selwyn Fund. Gift of Ralph Landau.

early days was abroad, and one of their first successes was an improved method of producing terephthalic acid—the main raw ingredient in polyester fiber—by bromine-assisted oxidation of paraxylene. Worldwide rights for this process were purchased by Standard Oil of Indiana (now Amoco). Another triumph was an improved process for producing propylene oxide, a substance used in polyurethane foams and in rigid poly-

mers; in this case the partner in the new corporation, called Oxirane, was Atlantic Richfield (Arco).

More recently, as a faculty member of the economics department at Stanford University and a fellow at Harvard's Kennedy School of Government, Landau has focused on understanding the political and economic environment necessary to encourage technological innovation—the lifeblood of a successful economy.

11. Human and Natural Environmental Concerns

The chemical industry has made many contributions to humanity's quality of life, as can be seen, for example, in the histories of chemists who invented processes to extract aluminum from the earth and gases from the air and who created new substances of all kinds, including those that make possible modern electronic communications and life-saving pharmaceuticals. The industry has also created problems that demand solution, and scientists have in the past alerted the public to these dangers and proposed appropriate solutions. In the future chemistry and chemical engineering will strive to create new processes that do not cause environmental problems.

CHARLES F. CHANDLER (1836–1925)

Charles F. Chandler was a pioneer in his commitment to both the urban and natural environments. His work with New York City's Metropolitan Board of Health between 1867 and 1883 provided a model for health and environmental laws and regulatory agencies nationwide by monitoring food and drugs, providing free vaccinations, ensuring the safety of milk supplies, bringing clean water into the city, and enacting building codes with adequate provisions for indoor plumbing (which he personally designed with appropriate trapping systems). Health problems were created not only by large numbers of people living in unprecedented proximity, but also by a burgeoning, if still chemically primitive, industry: the "nuisances" of noxious gases and acids discharged in sludge; dangerous products (e.g., kerosene that contained explosive naphtha fractions); and adulterated food, beverages, and cosmetics.

Chandler, the eldest son of an old and moderately successful New England family, prepared for his multifaceted career by attending Harvard University, after which he sailed for Germany. There he studied for his doctorate with Friedrich Wöhler (see Wöhler, Chapter 6) and Heinrich Rose, another former Berzelius student who followed up his teacher's interest in mineralogy and analytical chemistry. Charles's younger brother William, who also became a prominent figure in American chemistry, was instead educated in the United States—at Union College, Columbia School of Mines, and Hamilton College, from which he received his doctorate. At Union and at the School of Mines, William studied and worked with Charles, who established modern chemical studies at both institutions. At a time when salaries and fees were very low for academic chemists and holding multiple appointments was common, Charles conveyed the excitement of chemistry to students at Columbia College, the New York

■ Charles F. Chandler with a wash bottle in hand.
Courtesy Chandler Museum, Columbia University.

■ "A Proper Reception for King Cholera." Charles Chandler, with torch held high in this 1873 newspaper cartoon, led the drive for better sanitation in New York City as president of the Board of Health. Courtesy Chandler Museum, Columbia University.

College of Pharmacy, and the College of Physicians and Surgeons as well—the three of which eventually joined to become Columbia University.

Although Charles was preeminent as a sanitarian, he became equally famous as a consulting industrial chemist, a lucrative complement to his academic income. His range of chemical interests was extraordinary and included such topics as sugar, petroleum, illuminating gas, photographic materials, aniline dyes, and electrochemistry as well as the analysis of water and minerals more typical of a contemporary consulting chemist. He regarded industry as an exciting career opportunity for his many students, unlike some fellow academics, whose attitudes were more "ivory tower."

Charles Chandler was also a genius at organizing the American chemical community. He served as chairman of the chemists who gathered in 1874 at the grave of Joseph Priestley in Northumberland, Pennsylvania, to celebrate the centennial of the discovery of oxygen. At this meeting the notion of founding a national chemical society was conceived. The idea was publicized in *American Chemist*, a journal that he and his brother—by then chairman of chemistry at Lehigh University—edited from 1870 to 1877. Charles was a prime mover in founding the American Chemical Society in 1876 and its *Journal of the American Chemical Society*, which succeeded *American Chemist*. Charles served as president of the ACS in 1881 and 1889. He was elected the second chairman of the New York (later American) Section of the Society of Chemical Industry and was the first American to be chosen president of the London-based parent society (1899–1900). He was also an organizer and the first president (1898–1900) of The Chemists' Club, a club whose goal was to foster a social and professional identity in the chemical community associated with the nascent American chemical industry, then centered in New York City.

ELLEN SWALLOW RICHARDS (1842–1911)

The most prominent female American chemist of the nineteenth century, Ellen Swallow Richards, was another pioneer in sanitary engineering and a founder of home economics in the United States.

Swallow, the daughter of an old but relatively poor New England family, was taught that a good education was important. After years of teaching school, tutoring, and cleaning houses, she earned enough money to attend one of the new women's colleges. With the $300 she had saved, she entered Vassar College in 1868 as a special student and graduated two years later. At Vassar she was attracted to astronomy and chemistry. Upon graduation she applied for positions with various industrial chemists, but was turned down in all cases. At the suggestion of one of these chemists, however, she applied and was accepted as a special student at the Massachusetts Institute of Technology—the first woman in America to be accepted by a scientific school. Three years later she received a second bachelor's degree, a B.S. from MIT, as well as a master's degree from Vassar, to which she had submitted a thesis on the chemical analysis of an iron ore. She then continued at MIT with hopes of earning a doctorate, but MIT was not to award its first doctorate to a woman until 1886.

In 1875 Swallow married Robert Hallowell Richards, chairman of the mining engineering department. Supported in her ambitions by her husband, Ellen Swallow Richards volunteered her services as well as one thousand dollars annually to further women's scientific education at MIT. Through her efforts the Women's Laboratory was established in 1876, and in 1879 she was recognized as an assistant instructor—without pay—for teaching the curriculum in chemical analysis, industrial chemistry, mineralogy, and applied biology. The laboratory was closed in 1883 after MIT began awarding degrees to women on a regular basis, and there was no more need for a special track.

Coincidentally, in the same year MIT opened the nation's first laboratory of sanitary chemistry, headed by William R. Nichols, one of Richards's former professors, and she was appointed as an instructor. In 1887, at the request of the Massachusetts State Board of Health, the laboratory undertook a survey of the quality of the inland bodies of water of Massachusetts, many of which were already polluted with industrial waste and municipal sewage. Richards and her assistants performed the actual laboratory work and kept records. The scale of the survey was unprecedented: It led to the first state water-quality standards in the nation and the first modern municipal sewage treatment plant, in Lowell, Massachusetts. From 1887 to 1897 Richards served as official water analyst for the State Board of Health while continuing as an instructor at MIT—the rank she held at her death. She and her colleague A. G. Woodman wrote a classic text in the field of sanitary engineering: *Air, Water, and Food from a Sanitary Standpoint* (1900).

From her days at the Women's Laboratory, Richards was very concerned to apply scientific principles to domestic topics—good nutrition, pure foods, proper clothing, physical fitness, sanitation, and efficient practices that would allow women more time for pursuits other than cooking and cleaning. In 1882 she published *The Chemistry of Cooking and Cleaning: A Manual for House-keepers*. By setting up model kitchens open to the public, establishing programs of study, and organizing conferences, Richards campaigned tirelessly for the new discipline of home economics. Growing out of several summer conferences held at Lake Placid, New York, the American Home Economics Association was formed in 1908 with Richards as its first president.

■ Ellen Swallow about 1858.
Courtesy MIT Museum.

■ Ellen Swallow Richards (left rear)
with women students at MIT in 1888.
Courtesy MIT Museum.

■ Ellen Swallow Richards with an assistant in 1901 gathering samples from Jamaica Pond, near Boston, in connection with her systematic examination of the water supplies of the state of Massachusetts. Courtesy Sophia Smith Collection, Smith College.

ALICE HAMILTON (1869–1970)

Alice Hamilton was a founder of industrial toxicology in the United States. Her long career investigating the hazards to which workers were exposed—especially from contact with poisonous materials—earned her great gratitude and respect, even from many of the industrialists whose businesses she studied. As Bradley Dewey, president of Dewey and Almy Chemical Company, wrote in 1934 to the technical director of a firm that sold solvents:

I don't know what your Company is feeling as of today about the work of Dr. Alice Hamilton on benzol [benzene] poisoning. I know that back in the old days some of your boys used to think that she was a plain nuisance and just picking on you for luck. But I have a hunch that as you have learned more about the subject, men like your good self have grown to realize the debt that society owes her for her crusade. I am pretty sure that she has saved the lives of a great many girls in can-making plants and I would hate to think that you didn't agree with me.

> Dewey to S. P. Miller, Feb. 9, 1933, Alice Hamilton papers, no. 40, Schlesinger Library, Radcliffe College, quoted in Barbara Sicherman, *Alice Hamilton: A Life in Letters* (Cambridge, Mass.: Harvard University Press, 1984).

Hamilton was one of four daughters and a son born to one of the founding families of Fort Wayne, Indiana. Her sister Edith became famous as the author of *The Greek Way* (1930) and other works about classical culture, after a career as the headmistress of a girls' school in Baltimore. Alice planned to become a medical doctor. After attending a girls' boarding school that gave scant attention to science, she spent a summer being tutored in chemistry and physics before entering the University of Michigan Medical School. (At that time U.S. medical schools accepted students directly from high school.) At Michigan she became fascinated with the subject of pathology and decided to become a research scientist rather than enter clinical practice. After completing her medical training, she returned briefly

to the University of Michigan for graduate studies, but soon she and her sister Edith set out for Germany to pursue their respective fields—bacteriology and classics. Unlike their male counterparts, the sisters were not welcomed into the German universities. Robert Koch, a founder of bacteriology, and Paul Ehrlich rejected Alice Hamilton's request to work with them in Berlin, but she was well received by Ehrlich's former colleagues in Frankfurt, including his cousin, Carl Weigert (see Ehrlich, Chapter 7). Upon her return to the United States, Hamilton became a research assistant at Johns Hopkins Medical School, where she worked mainly with Simon Flexner, a pathologist who went on to head the Rockefeller Institute in New York.

In 1897 Hamilton accepted an appointment as professor of pathology at the Women's Medical School of Northwestern University, which was dissolved soon after. She later worked as a bacteriologist at Chicago's Memorial Institute for Infectious Diseases. In Chicago she lived for many years at Jane Addams' Hull House, the most famous of the settlement houses founded by churches and universities at the dawn of the twentieth century. Settlement houses, staffed by idealistic college graduates, offered help to immigrants and other poor people who lived and worked in congested and run-down inner cities. Among the projects she carried out at Hull House, Hamilton applied her medical expertise to finding the causes for the high incidence of typhoid fever and tuberculosis in the surrounding community. In the tuberculosis study she identified bad working conditions as one of the factors that weakened the resistance of poor immigrants to the disease.

Because of Hamilton's public health experience, in 1908 the governor of Illinois appointed her to the Illinois Commission on Occupational Diseases. The commission decided to conduct a broad survey of industrially related diseases in Illinois—a groundbreaking study, which Hamilton agreed to oversee. From 1911 to 1920 she served as a special investigator for the

■ Alice Hamilton in 1919 at the time of her appointment to Harvard. Courtesy Schlesinger Library, Radcliffe College.

federal Bureau (later Department) of Labor. A landmark study done by Hamilton while in this position concerned the manufacture of white lead and lead oxide, substances that were then commonly used as pigments in the paint industry, and she made recommendations for safer working conditions. (The danger of lead poisoning among members of the general population—especially children—had not yet been recognized.) Among her other famous studies was her work investigating the poisonous effects on workers of manufacturing explosives, a study undertaken during World War I—a war she opposed on pacifist grounds—at the request of the National Research Council.

In 1919, as the leading expert in the field of industrial medicine, Hamilton was appointed assistant professor at Harvard Medical School, whose faculty and student body were all male. She was in fact the first woman professor in any field in the entire university. The mutually agreeable plan was for her to be in residence six months of each year; the rest of the time was to be spent on her surveys. In part because of this unusual arrangement, when she retired in 1935, she was still an assistant professor.

Over the years Hamilton played a prominent role in turning the attention of government and industry to the poisonous effects of aniline dyes, carbon monoxide, mercury, tetraethyl lead, radium (in wristwatch dials among other uses), benzene, the chemicals in storage batteries, and carbon disulfide and hydrogen sulfide gases created in the manufacture of viscose rayon.

Throughout her long life Hamilton maintained an active concern for international affairs and individual civil liberties. She was a member of the League of Nations Health Committee from 1924 to 1930. Because of the threat to humanity posed by the Nazis, she supported the entry of the United States into World War II—in contrast to her stance on World War I. Because she often publicly supported the right of people to hold and express unpopular views, she attracted the suspicion of authorities, and her activities were followed by the Federal Bureau of Investigation, even when she was in her nineties.

■ Alice Hamilton at the age of twenty-four, the year she graduated from medical school. Courtesy Michigan Historical Collections, Bentley Historical Library, University of Michigan.

■ Frederick Gardner Cottrell (right), the inventor of electrostatic precipitation and chief metallurgist of the Bureau of Mines when this photo was taken in 1916, and Walter A. Schmidt, president of Western Precipitation Company. Courtesy Research Corporation.

FREDERICK GARDNER COTTRELL (1877–1948)

While a professor at the University of California, Berkeley, Frederick Cottrell entered the pollution cleanup business because DuPont wanted to eliminate a problem in a process designed to manufacture sulfuric acid. DuPont hired Cottrell in 1906 as a consultant to its facility at Pinole, twenty miles north of Berkeley. The contact process used to create the acid was producing arsenic that poisoned the catalyst, so Cottrell determined that centrifuging the arsenic-contaminated sulfuric acid mists would remove the arsenic. Then came the problem of precipitating the purified mist—the innovation that became known as "Cottrellizing." He experimented with passing an electric charge to the mist globules, which then migrated to the oppositely charged electrode, where they could be collected.

In 1907 Cottrell applied electrostatics to a different process: A successful court suit filed by Solano County, California, against the Selby Smelting and Lead Company required the smelter to clean up its sulfurous smoke. The lead particles were filtered out by a "baghouse" fitted with two thousand woolen bags, each thirty feet long, through which the dust-laden gases were blown. Cottrell designed a precipitator to recover the sulfuric acid—again from a mist—used in dissolving gold from the gold-and-silver alloy found in the lead. Cottrell later installed similar equipment at a copper smelter and a cement factory and developed a related electrostatic process for de-emulsifying oil. The fame of these operations spread, and a presentation by Cottrell at the 1910 American Chemical Society meeting in San Francisco drove home to the world that a major process for cleaning up the air was on the market.

Cottrell, born in California, graduated from Berkeley in 1896 and did one year of graduate work there; next he taught high school in Oakland for three years, then journeyed to Europe to study first with Jacobus Henricus van't Hoff (see van't Hoff, Chapter 6) in Berlin and later with Wilhelm Ostwald in Leipzig, where he received his doctorate. He was on the faculty at Berkeley from 1903 to 1911, where he conducted his earliest ventures into electrostatic precipitation. In 1911 he resigned from the university to join the U.S. Bureau of Mines, of which he eventually became director in 1919. There he worked in the World War I programs to develop processes to fix nitrogen for explosives (since the United States did not have any plants that used the Haber-Bosch process [see Haber, Chapter 2] at the time) and to distill helium from air for lighter-than-air craft. From 1922 to 1930 he served as the director of the Fixed Nitrogen Research Laboratory in the Department of Agriculture, which succeeded in developing a good catalyst for a Haber-type process. As director he was also responsible for recommending what to do with the nitrogen plant erected by the government at Muscle Shoals on the Tennessee River during World War I, which was converted from explosives to fertilizer manufacturing after the war. Cottrell's recommendation that the government continue to operate it as an experimental facility was ultimately incorporated in the plans for the Tennessee Valley Authority.

Next to the electrostatic removal of particles from smokestack gases, Cottrell is probably best remembered for his creation of the Research Corporation in 1912. This foundation was set up to receive income from his patents and the patents of other public-spirited inventors and to distribute these funds to university researchers in the physical sciences as seed money—which bore fruit in many instances. Before World War II, when federal support for scientific research was slight, the Research Corporation provided much-needed funds for Ernest Lawrence's development of the cyclotron, R. H. Goddard's experiments with rockets, the processes for volume production of vitamins A and B_1, and some of Robert Burns Woodward's early organic syntheses of complex organic molecules like the drug reserpine. The foundation continues to underwrite scientific research that might not otherwise gain support—especially projects undertaken by faculty members at small colleges.

■ Frederick Gardner Cottrell in 1920.
Courtesy Research Corporation.

■ Rachel Carson watching hawks at Hawk Mountain, near Reading, Pennsylvania, in 1945. Copyright Shirley Briggs. By permission of Rachel Carson History Project.

RACHEL CARSON (1907–1964)

Silent Spring, written in 1962 by Rachel Carson, brought to public attention the results of indiscriminate use of DDT (dichlorodiphenyltrichloroethane) and other pesticides. Carson also criticized industrial society for abusing the natural environment and failing to recognize the threat to industry's own existence when natural processes are seriously disturbed. Far more than earlier calls to use modern technology responsibly, *Silent Spring* launched a revolution in attitudes at all levels of society—from schoolchildren to government and in-

dustrial leaders. Carson's power did not stem from a charismatic personality, but lay in her scientific knowledge and poetic writing.

Born and raised in Springdale, Pennsylvania, near Pittsburgh, Carson witnessed how coal mining was despoiling the rural setting she loved. Already a devoted writer, as a child she published several stories in the children's magazine *St. Nicholas*. At Pennsylvania College for Women, now Chatham College, where she had been recruited as a scholarship student for her proven writing ability, she changed her major to biology—to the chagrin of some faculty members. After graduation she held a summer study fellowship at the Marine Biological Laboratory at Woods Hole, Massachusetts. There she first became acquainted with the ocean, which later became the topic of several of her best-selling books. She then entered Johns Hopkins University and completed a master's degree in marine zoology while serving as a teaching assistant and part-time instructor in biology at Johns Hopkins and the University of Maryland. When her father died, Carson became the sole support of her mother, who soon after had to raise two grandchildren when her other daughter died in 1937.

With the nation still in the Depression, Carson secured her first position with the U.S. Bureau of Fisheries (now the Fish and Wildlife Service of the Interior Department) in a temporary job writing radio scripts on marine life. From 1936 to 1952 she was a full-time employee of the Fish and Wildlife Service, moving into positions that drew on her abilities as a writer and editor; she was finally appointed editor-in-chief of the information division. Meanwhile, she wrote her own articles and books in the evenings and on weekends, publishing *Under the Sea Wind* (1941) and *The Sea Around Us* (1951), her first bestseller, which enabled her to resign from the Fish and Wildlife Service and devote all her energies to writing. Next came *The Edge of the Sea* (1955) and "Help Your Child to Wonder"—a series of articles, later reprinted as a book, based on her experiences with her grandnephew, whom she adopted in 1957 when his mother died.

In writing *Silent Spring*, Carson was transformed from a beloved nature writer into a crusader, battling in her quiet way with powerful political and economic interests. Although DDT was popularly viewed as a miracle of modern technology—especially because it had been successfully used in World War II to kill fleas, mosquitoes, and other insects that can spread deadly diseases—biologists had begun to compile evidence of the rise of DDT-resistant strains of insect pests and of the harmful side effects of DDT on other species. The U.S. Department of Agriculture—and the manufacturers of DDT and similar pesticides—nonetheless continued to support strongly the use of these substances. But Carson's gripping accounts of ecological disasters were based on a meticulous search and use of scientific literature, and her conclusions were upheld by President John F. Kennedy's Science Advisory Committee, among other authorities. Carson, who was diagnosed with breast cancer in 1957, lived only two years after *Silent Spring* was published and witnessed just the beginning of the groundswell of support for her views.

Two programs indicate how far *Silent Spring* transformed the nation's consciousness. Earth Day was first celebrated in 1970, and in 1988 a Chemical Manufacturers Association program called "Responsible Care" was established to help the chemical industry improve its safe management of chemicals from manufacture to disposal. With this program CMA continues to build on its decades of concern and involvement with ecological issues.

■ Rachel Carson on dock at Woods Hole, Massachusetts, in 1951. Copyright Edwin Gray. By permission of Rachel Carson History Project.

JOHN E. FRANZ (1929–) AND MARINUS LOS (1933–)

In recent years the ability of chemical scientists to create new molecules and transform existing ones has greatly contributed to preserving our environment. Since the 1960s chemists have engineered new pesticides that break down more rapidly instead of accumulating in soil and water. Two chemists, John E. Franz of Monsanto and Marinus Los of American Cyanamid, devised new herbicides—necessary for ensuring an abundant food supply for a growing world population—that do not harm the environment.

From the age of ten Franz knew that he wanted to become a scientist. In thrall to chemistry he had to be persuaded by a friendly high school teacher to study physics, mathematics, and other subjects in order to prepare properly for college. At the University of Illinois (B.S., 1951) he focused on chemistry, but not to the exclusion of other subjects. He went on to complete a Ph.D. in organic chemistry at the University of Minnesota, and then joined Monsanto. After a dozen years he was transferred to the company's agricultural unit, where he had to teach himself plant physiology and biochemistry. Franz, in his new position, began work on a project that other investigators had dropped—finding a herbicide effective against both perennial and annual weeds. The earlier investigators knew that two compounds were weakly active in this regard, but after nine years of trying various molecular analogs, they failed to produce a more effective herbicide. Franz tried a similar line of research for a year and finally hypothesized—falsely, as it turned out—that the weakly effective compounds were metabolized in the plant into herbicidal agents. On the basis of this hypothesis he screened possible metabolites and found glyphosate, a broadly effective herbicide with no toxic effect on mammals, birds, fish, insects, and most bacteria. Only later was the mechanism and selectivity of the product (sold as "Roundup") correctly explained; the glyphosate inhibited the formation of an enzyme found only in plants.

Los discovered another class of herbicidal compounds that are not toxic to humans and animals—imidazolinones. He was born in the Netherlands but emigrated as a boy with his family to Great Britain, where he received most of his education, including bachelor's and doctoral degrees in chemistry from the University of Edinburgh. After a postdoctoral fellowship in Canada he joined the agricultural division of American Cyanamid, where he worked for his entire career, except for a brief return to the University of Edinburgh as a research professor. In the early 1970s, while screening thousands of existing American Cyanamid chemicals for substances that would regulate plant growth, Los noted that one seemed to have herbicidal potential, so he made modifications in its molecular structure to enhance this capability. The first imidazolinone was actually created later when Los was trying to reproduce a substance that had crystallized out of a solution of the original herbicidal agent prepared for testing on crops. The new herbicides, introduced in 1985, need only be applied in ounces per acre, a property that has drastically reduced the amount of herbicide used. These highly selective and effective compounds allow farmers to increase yields while helping to preserve our land, water, and wildlife.

■ Monsanto's John Franz holding a crystal of the isopropylamine salt of glyphosate, the active ingredient in Roundup. Courtesy Monsanto Company.

■ Marinus Los, director of crop science discovery at American Cyanamid's agricultural research laboratory in Princeton, holds a molecular model of one of the imidazolinone herbicides he discovered in the 1970s. Courtesy American Cyanamid Company.

JAMES ROTH (1925–)

The accomplishments of James Roth display how catalytic processes similar to those described in Chapter 8 can be used to sculpt environmentally friendly molecules.

Roth came of age during World War II. Having completed high school and attended two years of college, at the age of eighteen Roth was serving as a navigator aboard a Navy vessel that landed Marines on Iwo Jima during World War II. After the war Roth completed his bachelor's degree in chemistry—earned with credits from three different colleges and universities—and a Ph.D. in physical chemistry from the University of Maryland. After a succession of positions, including a year or two as part owner of a small paint manufacturing company, Roth joined Monsanto in 1960. There he used platinum catalysis to solve an emerging environmental problem, the fouling of streams and lakes with detergents. Using a new platinum-based catalyst, he was able to create linear, or straight-chain, detergent molecules that microorganisms found digestible, as opposed to the earlier branched-chain molecules that were not biodegradable. Commercial detergent manufacturers quickly switched to the biodegradable type.

At Monsanto, Roth also participated in research on homogeneous catalysis, which involves soluble liquid-phase catalysts instead of the usual solid ones. The research resulted in a revolutionary process for making acetic acid from methanol. Carbon monoxide is added to methanol in the presence of a catalyst, instead of oxidizing ethylene, which is derived from depletable petroleum resources. This process quickly became the standard.

When Roth became a research director at Monsanto, his group's catalytic process was chosen to produce Roundup, the herbicide that Franz had discovered (see Franz, earlier in this chapter). In 1980, as founder and director of the Corporate Science Center at Air Products, dedicated to exploratory research, Roth entered yet another area in which catalytic processes apply—industrial gases.

■ A plant for making linear olefins for biodegradable detergents, Chocolate Bayou, Texas. Courtesy Monsanto Company.

171

■ James Roth. Courtesy James Roth.

Bibliography

For writings on scientists besides those listed under the individual chapter titles, please refer to General Histories and Overviews, Collected Biographies, and Oral Histories at the end of this bibliography.

1. FORERUNNERS

Robert Boyle and Marie Boas Hall. *Robert Boyle on Natural Philosophy: An Essay with Selections from His Writings.* Westport, Conn.: Greenwood Press, 1980 (reprint of 1965 edition, Indiana University Press).

Brenda Buchanan, editor. *Gunpowder: The History of an International Technology.* Bath, United Kingdom: Bath University Press, 1996.

Archibald Clow; Nan L. Clow. *The Chemical Revolution.* London: Gordon & Breach Science Publishers, 1992.

Arthur Donovan. *Antoine Lavoisier: Science, Administration and Revolution.* New York: Cambridge University Press, 1996.

Du Pont: The Autobiography of an American Enterprise. Wilmington, Del.: E. I. Du Pont de Nemours & Co., 1952. Distributed by Charles Scribner & Sons.

Henry Guerlac. *Lavoisier—The Crucial Year: The Background and Origin of His First Experiments on Combustion in 1772.* Ithaca, N.Y.: Cornell University Press, 1961.

Frederic L. Holmes. *Lavoisier and the Chemistry of Life: An Exploration of Scientific Discovery.* Madison: University of Wisconsin Press, 1985.

Michael Hunter, editor. *Robert Boyle, By Himself and His Friends: With a Fragment of William Woton's Lost* Life of Boyle. Brookfield, Vt.: William Pickering, 1994.

————. *Robert Boyle Reconsidered.* New York: Cambridge University Press, 1994.

Jean-Pierre Poirier. *Lavoisier: Chemist, Biologist, Economist.* Translated, with an introduction by Charles C. Gillispie. Philadelphia: University of Pennsylvania Press, 1996. Also available from the Chemical Heritage Foundation.

Joseph Priestley. *A Scientific Autobiography of Joseph Priestley, 1733–1804, Selected Scientific Correspondence.* Edited, with commentary by Robert E. Schofield. Cambridge, Mass.: MIT Press, 1966.

Rose-Mary Sargent. *The Diffident Naturalist: Robert Boyle and the Philosophy of Experiment.* Chicago: University of Chicago Press, 1995.

Truman Schwartz; John McEvoy, editors. *Motion toward Perfection: The Achievement of Joseph Priestley.* Boston: Skinner House Books, 1990.

2. THEORY AND PRODUCTION OF GASES

Maurice P. Crosland. *Gay-Lussac: Scientist and Bourgeois.* New York: Cambridge University Press, 1978.

Kenne Fant. *Alfred Nobel: A Biography.* Translated by Marianne Ruuth. New York: Arcade, 1993.

L. F. Haber. *The Poisonous Cloud: Chemical Warfare in the First World War.* New York: Oxford University Press, 1986.

Mikael Hard. *Machines Are Frozen Spirit: The Scientification of Refrigeration and Brewing in the 19th Century.* Frankfurt am Main: Campus Verlag; Boulder, Colorado: Westview, 1994. Includes information on Carl von Linde.

Dietrich Stoltzenberg. *Fritz Haber: Chemiker, Nobelpreisträger, Deutscher, Jude.* Weinheim, Germany/New York: VCH, 1994. English translation in press. Washington, D.C.: American Chemical Society and Chemical Heritage Foundation, 1998.

3. ELECTROCHEMISTRY AND ELECTROCHEMICAL INDUSTRIES

Edward G. Acheson. *A Pathfinder.* Port Huron, Mich.: Acheson Industries, 1965.

E. N. Brandt. *Growth Company: Dow Chemical's First Century.* East Lansing: Michigan State University Press, 1997.

Murray Campbell; Harrison Hatton. *Herbert H. Dow: Pioneer in Creative Chemistry.* New York: Appleton-Century-Crofts, 1951.

Geoffrey Cantor; David Gooding; Frank A. J. L. James. *Michael Faraday.* Atlantic Highlands, N.J.: Humanities Press, 1996.

Charles C. Carr. *Alcoa: An American Enterprise.* New York: Rinehart, 1952.

Elisabeth Crawford. *Arrhenius: From Ionic Theory to the Greenhouse Effect.* Canton, Mass.: Science History Publications, 1996.

Junius David Edwards. *Immortal Woodshed: The Story of the Inventor Who Brought Aluminum to America.* New York: Dodd, Mead, 1955. A biography of Charles Hall.

Margaret B. W. Graham; Bettye H. Pruitt. *R&D for Industry: A Century of Technical Innovation at Alcoa.* Cambridge: Cambridge University Press, 1990.

Harold Hartley. *Sir Humphry Davy.* London: Nelson, 1966.

David Knight. *Humphry Davy: Science and Power.* New York: Cambridge University Press, 1996.

Evan M. Melhado. *Jacob Berzelius: The Emergence of His Chemical System.* Madison: University of Wisconsin Press, 1981.

Evan M. Melhado; Tore Frängsmyr, editors. *Enlightenment Science in the Romantic Era.* New York: Cambridge University Press, 1992. Includes information on Jöns Jakob Berzelius.

Carol L. Moberg, editor. *The Beckman Symposium on Biomedical Instrumentation.* Fullerton, Calif.: Beckman Instruments, 1986.

George David Smith. *From Monopoly to Competition: The Transformations of Alcoa, 1888–1986.* Cambridge: Cambridge University Press, 1988.

Harrison Stephens. *Golden Past, Golden Future: The First Fifty Years of Beckman Instruments, Inc.* Claremont, Calif.: Claremont University Press, 1985.

John T. Stock; Mary Virginia Orna, editors. *Electrochemistry Past and Present.* Washington, D.C.: American Chemical Society, 1989.

Raymond Szymanowitz. *Edward Goodrich Acheson.* New York: Vantage Press, 1971.

Martha Moore Trescott. *The Rise of the American Electrochemicals Industry, 1880–1910.* Westport, Conn.: Greenwood Press, 1981.

Don Whitehead. *The Dow Story: The History of the Dow Chemical Company.* New York: McGraw-Hill, 1968.

L. Pearce Williams. *Michael Faraday.* New York: Da Capo, 1965.

4. THE PATH TO THE PERIODIC TABLE

John Bradley. *Before and After Cannizzaro: A Philosophical Commentary on the Development of the Atomic and Molecular Theories.* North Humberside, England: J. Bradley, 1992.

Sheldon Jerome Kopperl. "The Scientific Work of Theodore William Richards," Ph.D. diss., University of Wisconsin, 1970.

Mario A. Morselli. *Amedeo Avogadro: A Scientific Biography.* Boston: D. Reidel, 1984. Distributed by Kluwer Academic Publishers.

R. E. Oesper. "Robert Wilhelm Bunsen." *Journal of Chemical Education* 4 (1927), 431–439.

Elizabeth C. Patterson. *John Dalton and the Atomic Theory.* Garden City, N.Y.: Doubleday, 1970.

Alan J. Rocke. *Chemical Atomism in the Nineteenth Century.* Columbus: Ohio State University Press, 1982.

Arnold Thackray. *John Dalton: Critical Assessments of His Life and Science.* Cambridge, Mass.: Harvard University Press, 1972.

Morris William Travers. *A Life of Sir William Ramsay.* London: E. Arnold, 1956.

Jan W. van Spronsen. *The Periodic System of Chemical Elements: A History of the First Hundred Years.* New York: Elsevier, 1969.

Alexander Vucinich. "Mendeleev's Views on Science and Society." *Isis* 58 (1967), 342–351.

5. ATOMIC AND NUCLEAR STRUCTURE

T. E. Allibone. *Rutherford: The Father of Nuclear Energy.* Manchester: Manchester University Press, 1973.

Bernadette Bensaude-Vincent. "Star Scientists in a Nobelist Family: Irène and Frédéric Joliot-Curie." In *Creative Couples in the Sciences,* edited by H. M. Pycior, N. G. Slack, and P. G. Abir-Am. New Brunswick, N.J.: Rutgers University Press, 1996.

Ernst Berninger. *Otto Hahn.* Bonn: Inter Nationes, 1970.

Deborah Crawford. *Lise Meitner: Atomic Pioneer.* New York: Crown Publishers, 1969.

Eve Curie. *Madame Curie: A Biography.* Translated by Vincent Sheean. Reprinted, with a foreword by C. Stewart Gillmor, Norwalk, Conn.: Easton Press, 1989.

Otto Hahn. *My Life: The Autobiography of a Scientist.* Translated by Ernst Kaiser and Eithne Wilkins. New York: Herder and Herder, 1970.

Naomi Pasachoff. *Marie Curie and the Science of Radioactivity.* New York: Oxford University Press, 1996.

Rosalind Pflaum. *Grand Obsession: Madame Curie and Her World.* New York: Doubleday, 1989.

Susan Quinn. *Marie Curie: A Life.* Reading, Mass.: Addison Wesley Longman, 1995.

Glenn T. Seaborg. *A Chemist in the White House: From the Manhattan Project to the End of the Cold War.* Washington, D.C.: American Chemical Society, 1996.

———. *The Plutonium Story: Journals of Professor Glenn T. Seaborg, 1939–1946.* Columbus, Ohio: Battelle Press, 1994.

Glenn T. Seaborg, with Benjamin S. Loeb. *The Atomic Energy Commission under Nixon: Adjusting to Troubled Times.* New York: St. Martin's Press, 1993.

———. *Kennedy, Khrushchev, and the Test Ban.* Berkeley/Los Angeles: University of California Press, 1981.

———. *Stemming the Tide: Arms Control in the Johnson Years.* Lexington, Mass: Lexington Books, 1987.

Ruth Lewin Sime. *Lise Meitner: A Life in Physics.* Berkeley/Los Angeles: University of California Press, 1996.

Sir G. P. Thomson. *J. J. Thomson, Discoverer of the Electron.* Garden City, N.Y.: Anchor Books, 1966.

———. *J. J. Thomson and the Cavendish Laboratory in His Day.* Garden City, N.Y.: Doubleday, 1965.

David Wilson. *Rutherford, Simple Genius.* Cambridge, Mass.: MIT Press, 1983.

6. CHEMICAL SYNTHESIS, STRUCTURE, AND BONDING

Theodor Benfey. *From Vital Force to Structural Formulas.* Philadelphia: Chemical Heritage Foundation, 1992 (reprint of 1964 edition, Houghton Mifflin).

Theodor Benfey, editor. *Kekulé Centennial.* Washington, D.C.: American Chemical Society, 1966. On Kekulé's benzene theory and related topics.

William H. Brock. *Justus von Liebig: The Chemical Gatekeeper.* New York: Cambridge University Press, 1997.

Chemical Society of Great Britain. *The Life and Work of Professor William Henry Perkin.* London: Chemical Society, 1932.

Ted Goertzel; Ben Goertzel. *Linus Pauling: A Life in Science and Politics.* New York: Basic Books, 1995.

Thomas Hager. *Force of Nature: The Life of Linus Pauling.* New York: Simon & Schuster, 1995.

In Honor of Gilbert N. Lewis on his 70th Birthday. Berkeley/Los Angeles: University of California Press, 1945.

David E. Newton. *Linus Pauling: Scientist and Advocate.* New York: Facts on File, 1994. Makers of Modern Science series.

Linus Pauling. *Linus Pauling in His Own Words.* Edited by Barbara M. Marinacci. New York: Simon & Schuster, 1995.

Perkin Centenary, London: 100 Years of Synthetic Dyestuffs. Tetrahedron Supplement, 1. New York: Pergamon Press, 1958.

O. Bertrand Ramsay. *Stereochemistry.* Philadelphia: Heyden, 1981.

———, **editor.** *Van't Hoff–Le Bel Centennial.* Washington, D.C.: American Chemical Society, 1975. On stereochemistry.

Robert Scott Root-Bernstein. "The Ionists: Founding Physical Chemistry, 1872-1890." Ph.D. diss., Princeton University, 1980.

Albert Rosenfeld. *The Quintessence of Irving Langmuir.* Oxford/New York: Pergamon Press, 1966.

Margaret W. Rossiter. *The Emergence of Agricultural Science: Justus Liebig and the Americans, 1840–1880.* New Haven, Conn.: Yale University Press, 1975.

John W. Servos. *Physical Chemistry from Ostwald to Pauling: The Making of a Science in America.* Princeton, N.J.: Princeton University Press, 1990.

Virginia Veader Westervelt. *Incredible Man of Science.* New York: Messner, 1968. A biography of Irving Langmuir.

Howard J. White, Jr., editor. *Proceedings of the Perkin Centennial, 1856–1956: Commemorating the Discovery of Aniline Dyes.* Research Triangle Park, N.C.: American Association of Textile Chemists and Colorists, 1956.

John Wotiz, editor. *The Kekulé Riddle: A Challenge for Chemists and Psychologists.* Vienna, Ill.: Cache River Press, 1993.

7. PHARMACEUTICALS AND THE PATH TO BIOMOLECULES

Ernst Baumler. *Paul Ehrlich: Scientist for Life.* New York: Holmes & Meier, 1984.

G. M. Caroe. *William Henry Bragg, 1862–1942: Man and Scientist.* New York: Cambridge University Press, 1978.

Francis H. Crick. *What Mad Pursuit: A Personal View of Scientific Discovery.* New York: Basic Books, 1988.

Paul De Kruif. *Microbe Hunters.* New York: Harcourt, Brace, 1927. Reprinted with new introduction, San Diego: Harcourt Brace, 1996.

Carl Djerassi. *The Pill, Pygmy Chimps, and Degas' Horse.* New York: Basic Books, 1992.

———. *Steroids Made It Possible.* Profiles, Pathways, and Dreams, series editor Jeffrey I. Seeman. Washington, D.C.: American Chemical Society, 1990.

Guy Dodson, Jenny P. Glusker, and David Sayre, editors. *Structural Studies on Molecules of Biological Interest: A Volume in Honor of Dorothy Hodgkin.* Oxford: Clarendon Press; New York: Oxford University Press, 1981.

Henry Lowood, compiler. *William Henry Bragg and William Lawrence Bragg: A Bibliography of their Non-Technical Writings.* Berkeley: University of California, Berkeley, Office for the History of Science and Technology, 1978.

Martha Marquardt. *Paul Ehrlich.* New York: Schuman, 1951.

Robert Olby. *The Path to the Double Helix: The Discovery of DNA.* New York: Dover, 1994. Introduction by Francis Crick.

Anne Sayre. *Rosalind Franklin and DNA.* New York: W. W. Norton, 1978.

James D. Watson. *The Double Helix: A Personal Account of the Discovery of the Structure of DNA.* New York: W. W. Norton, 1980.

M. Weatherall. *In Search of a Cure: A History of Pharmaceutical Discovery.* New York: Oxford University Press, 1990.

8. PETROLEUM AND PETROCHEMICALS

American Chemical Society. *A National Historic Chemical Landmark: The Houdry Process for the Catalytic Conversion of Crude Petroleum to High-Octane Gasoline.* April 13, 1996. Booklet commemorating the designation of the Houdry process as a National Historic Chemical Landmark; available from the American Chemical Society.

C. G. Moseley. "Eugene Houdry, Catalytic Cracking, and World War II Aviation Gasoline." *Journal of Chemical Education* 61 (1984), 655–656.

Peter H. Spitz. *Petrochemicals: The Rise of an Industry.* New York: John Wiley & Sons, 1988.

9. PLASTICS AND OTHER POLYMERS

Stephen Fenichell. *Plastic: The Making of a Synthetic Century.* New York: Harper-Business, 1996.

Matthew Hermes. *Enough for One Lifetime: Wallace Carothers, Inventor of Nylon.* Washington, D.C.: American Chemical Society and Chemical Heritage Foundation, 1996.

G. B. Kauffman. "Wallace Hume Carothers and Nylon, the First Completely Synthetic Fiber." *Journal of Chemical Education* 65 (1988), 803–808.

Samuel P. Massie. "Behind 'Number, Please': The Story of W. Lincoln Hawkins." *Chemistry* 44 (16 Oct. 1971), 16.

———. "Henry Aaron Hill: The Second Mile." *Chemistry* 44 (11 Jan. 1971), 11.

Jeffrey L. Meikle. *American Plastic: A Cultural History.* New Brunswick, N.J.: Rutgers University Press, 1995.

Herbert Morawetz. *Polymers: The Origins and Growth of a Science.* New York: John Wiley & Sons, 1985.

Peter Morris. *Polymer Pioneers: A Popular History of the Science and Technology of Large Molecules.* Philadelphia: Chemical Heritage Foundation, 1986. Revised, 1990.

S. T. Mossman; Peter J. T. Morris, editors. *The Development of Plastics.* Cambridge: Royal Society of Chemistry, 1994.

Bernard E. Schaar. "Chance Favors the Prepared Mind. XII: Origins of the Plastics Industry." *Chemistry* 40 (19 Nov 1967), 19–20. A biographical sketch of Baekeland.

Raymond B. Seymour; Roger S. Porter, editors. *Manmade Fibers: Their Origin and Development.* New York: Elsevier Applied Science, 1993.

10. CHEMICAL ENGINEERING: CHEMISTRY SCALES UP FOR INDUSTRY

A Dollar to a Doughnut: Doc Lewis, as Remembered by his Former Students. New York: American Institute of Chemical Engineers, circa 1950.

Clark K. Colton. *Advances in Chemical Engineering: Research and Education.* New York: Academic Press, 1991.

William F. Furter, editor. *A Century of Chemical Engineering.* New York: Plenum Press, 1982.

———, **editor.** *History of Chemical Engineering.* Washington, D.C.: American Chemical Society, 1980.

E. J. Kahn. *The Problem-Solvers: A History of Arthur D. Little, Inc.* Boston: Little, Brown, 1986.

Ralph Landau. *Uncaging Animal Spirits: Essays on Engineering, Entrepreneurship, and Economics.* Edited by Martha Gottron. Cambridge, Mass.: MIT Press, 1994.

Arthur D. Little. *The Handwriting on the Wall: A Chemist's Interpretation.* Boston: Little, Brown, 1928.

Nicholas A. Peppas, editor. *One Hundred Years of Chemical Engineering.* Boston: Kluwer Academic Publishers, 1989.

Terry S. Reynolds. *75 Years of Progress: A History of the American Institute of Chemical Engineers, 1903–1983.* J. Charles Forman, Larry Resen, editors. New York: The Institute, 1983.

11. HUMAN AND NATURAL ENVIRONMENTAL CONCERNS

Frank Cameron. *Cottrell: Samaritan of Science.* Tucson: Research Corporation, 1993 (reprinted from 1952 edition).

Rachel Carson. *Silent Spring.* Boston: Houghton Mifflin, 1962. Reprinted with an introduction by Al Gore, 1994.

Robert Clarke. *Ellen Swallow: The Woman Who Founded Ecology.* Chicago: Follett, 1973.

Esther M. Douty. *America's First Woman Chemist: Ellen Richards.* New York: Messner, 1961.

Alice Hamilton. *Exploring the Dangerous Trades: The Autobiography of Alice Hamilton.* Boston: Northeastern University Press, 1985 (reprint of 1943 edition, Little, Brown).

Linda J. Lear. *Rachel Carson: Witness for Nature.* New York: Henry Holt, 1997.

Barbara Sicherman. *Alice Hamilton: A Life in Letters.* Cambridge, Mass.: Harvard University Press, 1984.

Eve Stwetka. *Rachel Carson.* New York: Franklin Watts, 1991.

GENERAL HISTORIES AND OVERVIEWS

Bernadette Bensaude-Vincent; Isabelle Stengers. *A History of Chemistry.* Cambridge, Mass.: Harvard University Press, 1997.

William H. Brock. *The Norton History of Chemistry.* New York: W. W. Norton, 1993.

Cathy Cobb; Harold Goldwhite. *Creations of Fire: Chemistry's Lively History from Alchemy to the Atomic Age.* New York: Plenum, 1995.

John Hudson. *The History of Chemistry.* New York: Chapman & Hall, 1992.

Aaron Ihde. *The Development of Modern Chemistry.* Rev. ed. New York: Dover, 1984 (first edition, 1964, Harper & Row).

David Knight. *Ideas in Chemistry: A History of the Science.* New Brunswick, N. J.: Rutgers University Press, 1992.

Mary Jo Nye. *Before Big Science: The Pursuit of Modern Chemistry and Physics, 1800–1940.* New York: Twayne Publishers (Simon & Schuster MacMillan), 1996.

Mary E. Weeks; Henry M. Leicester. *The Discovery of the Elements.* Seventh edition. Easton, Pa.: Journal of Chemical Education Press, 1968.

COLLECTED BIOGRAPHIES

American Men and Women of Science. New Providence, N.J.: R. R. Bowker, 1994 and earlier editions. Catalogue of prominent living scientists.

Biographical Memoirs of the National Academy of Sciences, 1877–. Washington, D.C.: National Academy Press, 1931.

Mary Ellen Bowden and John Kenly Smith. *American Chemical Enterprise.* Philadelphia: Chemical Heritage Foundation, 1994.

Dictionary of Scientific Biography. New York: Scribners, 1970–1990.

Eduard Farber, editor. *Great Chemists.* New York: Interscience Publishers, 1961.

Louise S. Grinstein; Rose K. Rose; Miriam H. Rafailovich, editors. *Women in Chemistry and Physics: A Bibliographic Sourcebook.* Westport, Conn./London: Greenwood Press, 1993.

Caroline L. Herzberg. *Women Scientists from Antiquity to the Present: An Index.* West Cornwall, Conn.: Locust Hill Press, 1986.

Laylin K. James, editor. *Nobel Laureates in Chemistry, 1901–1992.* Washington, D.C.: American Chemical Society; Philadelphia: Chemical Heritage Foundation, 1993.

G. Kass-Simon, Patricia Farnes, editors. *Women of Science: Righting the Record.* Bloomington: Indiana University Press, 1990.

James H. Kessler et al. *Distinguished African-American Scientists of the Twentieth Century.* Phoenix: Oryx Press, 1996.

Sharon McGrayne. *Nobel Prize Women in Science: Their Lives, Struggles, and Momentous Discoveries.* New York: Carol Publishing, 1993.

Wyndham D. Miles, editor. *American Chemists and Chemical Engineers.* Washington, D.C.: American Chemical Society, 1976.

Wyndham D. Miles; Robert F. Gould, editors. *American Chemists and Chemical Engineers, Vol. 2.* Guilford, Conn.: Gould Books, 1994.

Marilyn B. Ogilvie. *Women in Science: Antiquity through the Nineteenth Century.* Cambridge, Mass.: MIT Press, 1986.

Roy Porter, consulting editor. *The Biographical Dictionary of Scientists.* Second edition. New York: Oxford University Press, 1994.

Marelene and Geoffrey Rayner-Canham. *Women in Science.* Washington, D.C.: American Chemical Society; Philadelphia: Chemical Heritage Foundation, 1997.

Vivian Ovelton Sammons. *Blacks in Science and Medicine.* New York/London: Hemisphere Publishing, 1990.

ORAL HISTORIES

(All the following oral histories are on deposit at the Othmer Library of the Chemical Heritage Foundation. Those with limited access are so noted.)

Arnold Beckman. Oral history interview by Jeffrey Sturchio and Arnold Thackray, 23 April and 23 July 1985.

Carl Djerassi. Oral history interview by Jeffrey L. Sturchio and Arnold Thackray, 31 July 1985. (Limited access)

John E. Franz. Oral history interview by James J. Bohning, 29 November 1994. (Limited access)

N. Bruce Hannay. Oral history interview by James J. Bohning, 9 March 1995 and 28 December 1995.

Ralph Landau. Oral history interview by James J. Bohning, 18 December 1990. (Limited access)

Marinus Los. Oral history interview by James J. Bohning and Bernadette R. McNulty, 17 January 1995. (Limited access)

Donald Othmer. Oral history interview by James J. Bohning, 2 April and 11 June 1986, and 15 January 1987.

Linus Pauling. Oral history interview by Jeffrey L. Sturchio, 6 April 1987.

Roy Plunkett. Oral history interview by James Bohning, 14 April and 27 May 1986.

James Roth. Oral history interview by James J. Bohning, 23 January 1995. (Limited access)

John H. Sinfelt. Oral history interview by James J. Bohning, 21 February 1995.

Paul B. Weisz. Oral history interview by James J. Bohning, 27 March 1995. (Limited access)

Index

OTHER BOOKS AVAILABLE FROM CHF

BIOGRAPHIES & AUTOBIOGRAPHIES

Women in Chemistry
Their Changing Roles from Alchemical Times to the Mid-Twentieth Century

Marelene F. Rayner-Canham & Geoffrey W. Rayner-Canham

Though rarely noted, women have been active participants in the chemical sciences since the beginning of recorded history. This thought-provoking book goes beyond Marie Curie to profile over 50 women who made significant contributions to chemistry.

History of Modern Chemical Sciences Series
1998. 240 pp, illus, index
Paper, 6 × 9, ISBN 0-941901-27-0
$19.95

Robert Burns Woodward
Architect and Artist in the World of Molecules

Edited by Otto Theodor Benfey & Peter J. T. Morris

Robert Burns Woodward was the star of 20th-century organic chemistry. This volume presents Woodward's most celebrated papers and lectures, including the famous Cope lecture. Insightful commentaries and rarely seen photographs are also included.

History of Modern Chemical Sciences Series
2001. 497 pp, illus, bibl, index
Cloth, 8.5 × 11, ISBN 0-941901-25-4
$45.00

Nobel Laureates in Chemistry, 1901–1992

Edited by Laylin K. James

An authoritative and informative volume that examines the scientific achievements in chemistry for which the Nobel Prize has been awarded. Biographies of all 116 Nobel laureates explore their scientific achievements and their human side.

History of Modern Chemical Sciences Series
1993. 650 pp, profuse illus
Cloth, 6 × 9, ISBN 0-8412-2459-5
$69.95
Paper, 6 × 9, ISBN 0-8412-2690-3
$34.95

Polymer Pioneers
A Popular History of the Science and Technology of Large Molecules

Peter J. T. Morris

Biographical sketches of twelve pioneers, from Marcellin Berthollet and John Wesley Hyatt to Karl Ziegler and Giulio Natta.

1986. 88 pp, 80 illus, bibl
$12.00 ($8.00 each for 10 or more copies)

Everybody Wins!
A Life in Free Enterprise
Second edition

Gordon Cain

A case study in American business enterprise told through the life of an extraordinary man: Gordon Cain–a chemical engineer who became an entrepreneur when most of us are ready to retire–effected a turnaround in the commodity chemicals industry in his 70s. In his 80s he turned his interests to biotechnology, transforming a university-based start-up into a public company worth over one billion dollars.

Series in Innovation and Entrepreneurship
2001. xx + 354 pp, illus, index, apps
Paper, 6 × 9, ISBN 0-941901-28-9
$14.95

Enough for One Lifetime
Wallace Carothers, Inventor of Nylon

Matthew Hermes

The first full biography of the man whom Roger Adams called the greatest chemist in America. Hermes tells the story of Carothers's sudden dramatic research successes–the discoveries that led to neoprene and nylon–and his relentless slide into depression, alcohol, and suicide.

History of Modern Chemical Sciences Series
1996. 342 pp, illus
Cloth, 6 × 9, ISBN 0-8412-3331-4
$38.95

Arnold O. Beckman
One Hundred Years of Excellence

Arnold Thackray and Minor Myers, jr.

Arnold O. Beckman is a living legend: the blacksmith's son who grew up to play a pivotal role in the instrumentation revolution that has dramatically changed science, technology, and society. In 1934 he began one of the earliest high-tech startups in a garage, creating the first in a series of pathbreaking scientific instruments, the pH meter. His firm's pioneering technologies contributed not only to the life sciences but also to such secret World War II efforts as radar and the Manhattan Project. As head of Beckman Instruments he influenced the beginnings of Silicon Valley. Arnold Beckman's story is inseparable from that of the twentieth century–a very inspiring read.

Series in Innovation and Entrepreneurship, special folio volume
397 pp, profuse illus, index
Cloth, 9 × 11, ISBN 0-941901-23-8
$65.00, includes CD-ROM video

Lavoisier
Chemist, Biologist, Economist

**Jean-Pierre Poirier
Revised and translated, with a preface by Charles C. Gillispie**

Now available in English, this comprehensive biography covers Lavoisier's role in French economic thought and politics as well as in chemistry and treats Marie Lavoisier as a figure in her own right.

Chemical Sciences in Society Series, copublished with the University of Pennsylvania Press
1998. 544 pp, illus, bibl, notes, index
Paper, 6 × 9.25, ISBN 0-8122-1649-0
$19.95

Stalin's Captive
Nikolaus Riehl and the Soviet Race for the Bomb

Nikolaus Riehl and Frederick Seitz

"This historic book . . . presents Riehl's absorbing account of his key role in the production of pure uranium for . . . the Soviet nuclear bomb program. Written by uniquely qualified Frederick Seitz, an extraordinary and perceptive introduction."
—Glenn Seaborg, *Nobel laureate in chemistry*

History of Modern Chemical Sciences Series
1996. 218 pp, profuse illus
Cloth, 6 × 9, ISBN 0-8412-3310-1
$34.95

A Devotion to Their Science
Pioneer Women of Radioactivity

Marelene F. Rayner-Canham & Geoffrey W. Rayner-Canham, editors and senior authors

Biographical essays on 23 women who worked in atomic science during the first two decades of the 20th century, including Marie Curie, Lise Meitner, Irène Joliot-Curie, and a host of lesser-known women scientists whose life stories have never been told before.

A Devotion to Their Science provides new insights into the contribution of women to atomic science and dispels the myth that this field was essentially a male preserve.

Copublished with McGill-Queen's University Press.

1997. 280 pp, 6 × 9
Paper, ISBN 0-941901-15-7
$19.95
Cloth, ISBN 0-941901-16-5
$55.00
World rights, except Canada

OTHER BOOKS AVAILABLE FROM CHF

Forthcoming Biographies

COMING FALL 2002

Pharmaceutical Achievers

Mary Ellen Bowden, Amy Beth Crow, and Tracy Sullivan

This biographical collection highlights individuals who made outstanding achievements in the arenas of pharmaceuticals and biotechnology and concludes with a look at tomorrow's medicines.

2002. 187 pp, illus, bibl, index
Paper, 8.5 × 11, ISBN 0-941901-30-0
$30 (Teachers: Ship to your school's address and pay only $15.00!)

COMING 2003

Fritz Haber
Chemist, Nobel Laureate, German, Jew

Dietrich Stoltzenberg

Stoltzenberg's biography of Fritz Haber won major acclaim in Germany and received the German Chemical Society's author prize. An abridged version in English will soon be available, published by CHF.

Use the recent past of the twentieth century's central science as a guide to the future!

General Histories

The American Synthetic Rubber Program

Peter J. T. Morris
Foreword by Arnold Thackray

"A very readable account of the wartime project."—*Chemistry and Industry*

1989. 204 pp, notes, bibl, index
Cloth, 6.25 × 9.5, ISBN 0-8122-8205-1
Now only $9.95

GENERAL HISTORIES

A History of the International Chemical Industry
Second edition

Fred Aftalion

Aftalion presents an international perspective on the history of chemistry, integrating the story of chemical science with that of chemical industry, and emphasizing the developments of the 20th century.

This new edition also discusses 1990–2000, when major companies began selling off their divisions, seeking to specialize in a particular business. Aftalion explores the pitfalls these companies encountered as well as the successes of "contrarians"–those companies that remained broad and diversified. He uses BASF, Dow, and Bayer as examples of true contrarians.

2001. xx + 436 pp, illus, bibl, index
Paper, 6 × 9, ISBN 0-941901-29-7
$24.95

Pharmaceutical Innovation
Revolutionizing Human Health

Edited by Ralph Landau, Basil Achilladelis, and Alexander Scriabine

A wide-ranging look at an industry that is central to the health and welfare of humanity, this pioneering work documents how science has provided an astonishing array of medicines for coping with human ailments over the last 150 years. This work covers not only the history of the pharmaceutical industry in chronological terms but also industry leaders, economic influences, and the development of individual products.

Series in Innovation and Entrepreneurship
1999. 412 pp, illus, bibl, index, glossary
Cloth, 6 × 9, ISBN 0-941901-21-1
$44.95

Chemical Sciences in the Modern World

Edited by Seymour Mauskopf

"Chemistry is, in fact, the 20th century's central science. . . . *Chemical Sciences in the Modern World* . . . [demonstrates] how the insights of historical research can help chemists themselves, policy-makers, and the general public understand policy issues involving the chemical sciences."—*Science*

Chemical Sciences in Society Series,
copublished with the University of Pennsylvania Press
1994. 448 pp, 14 illus, notes, index
Cloth, 6.25 × 9.5, ISBN 0-8122-3156-2
$39.95

Measuring Mass
From Positive Rays to Proteins

Edited by Michael A. Grayson

Our knowledge of the elements received a tremendous boost nearly a hundred years ago when physicists began exploring the newly discovered phenomenon of "rays of positive electricity" in greater detail. This work led to the development of the first mass spectrographs, which were used to determine the exact mass and relative abundance of the elements and their isotopes. Even then, scientists recognized greater potential for the mass spectrograph.

Measuring Mass is the best available account of a technique that has had the most impact on scientific discovery in the twentieth century. It describes the diverse applications of mass spectrometry in present-day scientific endeavors, and it highlights the major events in the history of mass spectrometry.

2002. 160 pp, illus, index
Cloth, 8.5 × 11, ISBN 0-941901-31-9
$35.00

Private Science
Biotechnology and the Rise of the Molecular Sciences

Edited by Arnold Thackray

Private Science examines the relationships among corporations, universities, and national governments involved in biotechnological research.

Chemical Sciences in Society Series,
copublished with the University of Pennsylvania Press
A project of CHF's Biomolecular Sciences Initiative
1998. 304 pp
Cloth, 6.5 × 9.5, ISBN 0-8122-3428-6
$52.50

Inventing Polymer Science
Staudinger, Carothers, and the Emergence of Macromolecular Chemistry

Yasu Furukawa

Polymer science is central to material and intellectual life in the 20th century. Focusing on the work of two central figures, Hermann Staudinger and Wallace Carothers, Furukawa explores the history of modern polymer science by tracing its emergence from macromolecular chemistry—its true beginning.

Chemical Sciences in Society Series,
copublished with the University of Pennsylvania Press
1998. 416 pp, illus
Cloth, 6 × 9, ISBN 0-8122-3336-0
$49.95